The Drone Honey Bee

Authored By

Lovleen Marwaha
Department of Zoology
School of Bioengineering and Biosciences, Lovely Professional University
Punjab, India

The Drone Honey Bee

Author: Lovleen Marwaha

ISBN (Online): 978-981-5179-30-9

ISBN (Print): 978-981-5179-31-6

ISBN (Paperback): 978-981-5179-32-3

© 2023, Bentham Books imprint.

Published by Bentham Science Publishers Pte. Ltd. Singapore. All Rights Reserved.

First published in 2023.

BENTHAM SCIENCE PUBLISHERS LTD.
End User License Agreement (for non-institutional, personal use)

This is an agreement between you and Bentham Science Publishers Ltd. Please read this License Agreement carefully before using the ebook/echapter/ejournal (**"Work"**). Your use of the Work constitutes your agreement to the terms and conditions set forth in this License Agreement. If you do not agree to these terms and conditions then you should not use the Work.

Bentham Science Publishers agrees to grant you a non-exclusive, non-transferable limited license to use the Work subject to and in accordance with the following terms and conditions. This License Agreement is for non-library, personal use only. For a library / institutional / multi user license in respect of the Work, please contact: permission@benthamscience.net.

Usage Rules:

1. All rights reserved: The Work is the subject of copyright and Bentham Science Publishers either owns the Work (and the copyright in it) or is licensed to distribute the Work. You shall not copy, reproduce, modify, remove, delete, augment, add to, publish, transmit, sell, resell, create derivative works from, or in any way exploit the Work or make the Work available for others to do any of the same, in any form or by any means, in whole or in part, in each case without the prior written permission of Bentham Science Publishers, unless stated otherwise in this License Agreement.
2. You may download a copy of the Work on one occasion to one personal computer (including tablet, laptop, desktop, or other such devices). You may make one back-up copy of the Work to avoid losing it.
3. The unauthorised use or distribution of copyrighted or other proprietary content is illegal and could subject you to liability for substantial money damages. You will be liable for any damage resulting from your misuse of the Work or any violation of this License Agreement, including any infringement by you of copyrights or proprietary rights.

Disclaimer:

Bentham Science Publishers does not guarantee that the information in the Work is error-free, or warrant that it will meet your requirements or that access to the Work will be uninterrupted or error-free. The Work is provided "as is" without warranty of any kind, either express or implied or statutory, including, without limitation, implied warranties of merchantability and fitness for a particular purpose. The entire risk as to the results and performance of the Work is assumed by you. No responsibility is assumed by Bentham Science Publishers, its staff, editors and/or authors for any injury and/or damage to persons or property as a matter of products liability, negligence or otherwise, or from any use or operation of any methods, products instruction, advertisements or ideas contained in the Work.

Limitation of Liability:

In no event will Bentham Science Publishers, its staff, editors and/or authors, be liable for any damages, including, without limitation, special, incidental and/or consequential damages and/or damages for lost data and/or profits arising out of (whether directly or indirectly) the use or inability to use the Work. The entire liability of Bentham Science Publishers shall be limited to the amount actually paid by you for the Work.

General:

1. Any dispute or claim arising out of or in connection with this License Agreement or the Work (including non-contractual disputes or claims) will be governed by and construed in accordance with the laws of Singapore. Each party agrees that the courts of the state of Singapore shall have exclusive jurisdiction to settle any dispute or claim arising out of or in connection with this License Agreement or the Work (including non-contractual disputes or claims).
2. Your rights under this License Agreement will automatically terminate without notice and without the

need for a court order if at any point you breach any terms of this License Agreement. In no event will any delay or failure by Bentham Science Publishers in enforcing your compliance with this License Agreement constitute a waiver of any of its rights.
3. You acknowledge that you have read this License Agreement, and agree to be bound by its terms and conditions. To the extent that any other terms and conditions presented on any website of Bentham Science Publishers conflict with, or are inconsistent with, the terms and conditions set out in this License Agreement, you acknowledge that the terms and conditions set out in this License Agreement shall prevail.

Bentham Science Publishers Pte. Ltd.
80 Robinson Road #02-00
Singapore 068898
Singapore
Email: subscriptions@benthamscience.net

CONTENTS

FOREWORD i
PREFACE ii

CHAPTER 1 COMPREHENSIVE OVERVIEW OF APIS MELLIFERA DRONE
DEVELOPMENT, BIOLOGY, AND INTERACTION WITH THE QUEEN 1
    INTRODUCTION 1
    THE DEVELOPMENTAL SYNCHRONICITY OF DRONE HONEY BEES 2
    SOME FACTS ABOUT THE DIPLOID DRONES 3
    LIFE EXPECTANCY IN DRONE HONEY BEE 4
    DRONE POPULATION IN HONEY BEE COLONY 5
    FLIGHT ACTIVITY OF DRONE HONEY BEES 6
    DRIFT AND ORIENTATION OF DRONES' FLIGHT TO THE HIVE 7
    REPRODUCTIVE SYSTEM OF THE DRONE HONEY BEES 8
        Testes 8
        Vas Deferens 9
        Seminal Vesicle 9
        Mucous Glands 9
        Ejaculatory Duct 9
    MATING 9
    DRONE AGING 10
    CONCLUSION 10
    REFERENCES 11

CHAPTER 2 THE DRONE HONEY BEE MORPHOMETRIC CHARACTER 18
    INTRODUCTION 18
    MORPHOMETRIC CHARACTERS 20
    DRONE'S REPRODUCTIVE SYSTEM METRIC REVIEW 24
    CONCLUSION 25
    REFERENCES 25

CHAPTER 3 THE DEVELOPMENT OF THE DRONE HONEY BEES: THE
PARTHENOGENESIS 28
    INTRODUCTION 28
    DRONE DEVELOPMENT AND THE NEED FOR A PROTEIN-RICH DIET 31
    INFLUENCE OF ENVIRONMENTAL STRESSORS ON DRONE DEVELOPMENT ... 32
    GENERAL DRONE DEVELOPMENT 35
        Drone Honey Bee Egg Stage 35
        Larval Stage of the Drone Honey Bee 36
        The Pupal Development in the Drone Honey Bee 38
        The Adult Drone Honey Bee 38
        The life Span of Adult 38
    THE TEMPERATURE PREFERENCE OF ADULT DRONES 41
    FACTORS INFLUENCING DRONE CELLS AND LIFE SPAN 43
    CONCLUSION 44
    REFERENCES 44

CHAPTER 4 THE PHEROMONAL PROFILE OF THE DRONE HONEY BEES APIS
MELLIFERA: THE VOLATILE MESSENGERS 53
    INTRODUCTION 53
    THE PHEROMONAL COMMUNICATION IN THE COLONY 54
    THE DRONE MANDIBULAR GLAND PHEROMONES 55

DETECTION OF PHEROMONES BY DRONE HONEY BEES	57
DRONE CONGREGATION AREA	58
The Drone's Attraction Toward to Queen	60
CONCLUSION	61
REFERENCES	61
CHAPTER 5 THE MATING AND REPRODUCTION IN APIS MELLIFERA: THE ROLE OF DRONE HONEY BEE	65
INTRODUCTION	65
DRONE CONGREGATION AREA	67
DRONE FLIGHTS BEFORE MATING	69
THE DRONE'S REPRODUCTIVE POTENTIAL	69
CONCLUSION	73
REFERENCES	73
CHAPTER 6 ARTIFICIAL METHODS OF DRONE REARING IN APIS MELLIFERA AND THE ROLE OF DRONES IN QUALITY IMPROVEMENT	84
INTRODUCTION	84
GENOMIC CONTRIBUTION OF DRONES TO QUALITY IMPROVEMENT	87
CONCLUSION	89
REFERENCES	89
SUBJECT INDEX	93

FOREWORD

Honey bees play an important role in the sustainability of our agro-ecosystem. Bees not only provide pollination services but also provide hive products and thereby act as a crucial source of income for the masses. A thorough understanding of the biology of honey bee queens, workers and drones and the factors affecting their quality and performance helps in improving the productivity of honey bee colonies. A fertile quality queen bee is essential for the long-term survival and growth of a honey bee colony. Further, the productivity of the honey bee colony is also highly dependent on the quality of the queen bee. Therefore, the rearing of queen bees from the most productive honey bee colonies having the desired traits leads to the path to stock improvement. Drones in the honey bee colonies contribute by fertilizing the queen bees and thereby enabling them to lay fertilized eggs resulting in a huge population of worker bees. Therefore, the rearing of a large number of good quality drones from the selected best-performing honey bee colonies is equally important in the bee breeding program. However, for rearing good-quality drone bees, addressing the nutritional requirements of drones is also very important. Understanding the drone maturity, synchronization of drone maturity and queen maturity, mating flight, drone congregation area, and weather conditions suitable for queen bee mating success help us in better planning and execution of bee breeding. This book provides very good information on drone development, pheromonal communication, and the reproductive system of drones. The chapter on the mating and reproduction in *Apis mellifera* provides detailed information regarding drone congregation area and reproductive potential of drones and the factors which affect these parameters. This book also throws light on the significant role drones play in stock quality improvement. This book will provide useful information to the students and researchers. My best wishes to the author.

Jaspal Singh
Principal Entomologist Department of Entomology
Punjab Agricultural University
Ludhiana -141 004, India

PREFACE

'The Drone Honey Bee' has been written especially for B.Sc., M.Sc. and Ph.D. students, highlighting various aspects of the drone honey bee's life cycle. Books on honey bees are easily available in the market which gives immense knowledge about colony organization, different castes of the colony, communication in the colony, the productivity of the colony, genomics, proteomics, and others. But, drone honey bee is very less explored in comparison to other castes, being not directly involved in colony productivity and hence in economic benefits. However, drones can be used in the bee quality improvement of the colony. Drones can be reared artificially in the colony or outside the colony up to certain stages that can be larval, pupal or adult. The drone pupae are good protein supplements for human consumption. The current book elucidates the available details of the specific caste of honey bees.

The present book elaborates on general introduction, morphology, development, pheromonal profile, mating, reproduction, artificial drone development and genomic contribution of drones in colony improvement. The drone honey bee provides patrilineal genomic contributions to the honey bee colony that influence colony productivity, colonial behavior, adaptability, and others. Although the prevalence of drones is seasonal and as per the availability of food resources, the specific caste is an integral part of the colony with a chief role in mating and thermal regulation. The current book highlights information about drone honey bees as per the availability of literature.

Regards,

Lovleen Marwaha
Department of Zoology
School of Bioengineering and Biosciences, Lovely Professional University
Punjab,
India

CHAPTER 1

Comprehensive Overview of Apis mellifera Drone Development, Biology, and Interaction with The Queen

Abstract: The male honey bees, the reproductive caste of the colony, develop through haploid/diploid parthenogenesis. The drones develop from haploid/ diploid unfertilized eggs produced by parthenogenesis or from diploid fertilized eggs having identical sex alleles, formed after sexual reproduction, with more probability when the queen honey bee mates with the drones of the same hives. Therefore, two types of drone honey bees, based on ploidy, are common in colonies, *e.g.* haploid or/and diploid. The number of drone honey bees staying in the colony varies according to protein resources and the strength of the worker honey bees. Generally, the haploid drone eggs/larvae laid by workers are removed by the nurse bee due to cannibalism. The above-mentioned eggs/larvae are marked with certain specific hormones that act as markers for cannabalic removal of the same.

Further, the development of drones is influenced by colony temperature; hence overall development can be completed within 24-25 days. The purpose of drone life is to produce sperm and mate with the queen. The queen attracts the drone's honey bees toward herself with pheromones 9-ODA, 9-HDA and 10 HDA. The drone number and fertility depend upon the colony's environmental conditions, genomic possession and available food in the colony. The specific chapter provides deep insight into the development of drones, the biology of drones, the reproductive system, and the mating behaviour of particular castes. Subsequent chapters highlight morphometric characteristics of drones, development, mating, reproduction and artificial drone production.

Keywords: Haploid and Diploid Drones, Parthenogenesis, Developmental Synchronicity.

INTRODUCTION

The drone honey bees perform the function of mating and temperature regulation. Further, the concerned caste does not forage, maintain the hive, defend the colony or perform other functions. During the nuptial flight, the polyandrous honey bee queen usually mates with 6-17 drones (Peer *et al.*, 1956; Renner and Baumann, 1964; Adams *et al.*, 1977; Santomauro *et al.*, 2004), and post-mating death of drones is inevitable (Witherell, 1956). Unfortunately, the procurable scientific literature for drone honey bees is not vastly explored. Therefore, limited information is available on drones' contribution to agricultural pollination, apicultural production, or the protection of colonies. Further, drones can enhance honey bee colonial productivity, can make colony disease and swarm-resistant, can control overall behaviour and organization through genomic contribution, and others.

THE DEVELOPMENTAL SYNCHRONICITY OF DRONE HONEY BEES

The haploid drone honey bees usually carry maternal inheritance due to their formation from unfertilized eggs laid in drone/worker wax cells by the queen or from a haploid egg laid by the pseudo queen or egg-laying honey bee workers in queen-righted or queen-less colonies(Kerr, 1974a,b; Herrmann *et al.*, 2005; Brutscher *et al.*, 2019). The haplodiploid sex-determination mechanism is well exemplified in eusocial insect honey bees.

Even diploid drones can develop from fertilized eggs in case of the queen mates with drones of the same colony (Page and Laidlaw, 1985). Such diploid drones can carry inheritance from both maternal and paternal sides. Furthermore, egg-laying workers can lay diploid eggs, with two sets of chromosomes coming from one polar body and an ovum. The specific process is known as thelytoky, a type of parthenogenesis. Other workers sense such diploid eggs through coated pheromones; therefore, such diploid drones are eaten by workers within a few hours after the eggs hatch, which highlights the phenomenon of cannibalism in honey bee colonies (Woyke, 1965).

The developmental duration of drone honey bees varies according to temperature. The temperature is different in the hive's centre and periphery, so the drone's development varies. For the development of drones from egg to adult, about 24 days are required (Jay, 1963), whereas, in the peripheral areas of the hive, more time is usually needed, which could be up to 25 days (Fukuda and Ohtani, 1977). In other words, drone development is correlated with brood nest temperature variation (Free, 1967; Jay, 1963; Fukuda and Ohtani, 1977; Santomauro *et al.*, 2004).

For the general development of drones, about three days are required for the egg to hatch, six days for larval development, and 15 days for the pupal phase.

SOME FACTS ABOUT THE DIPLOID DRONES

Haploid drones develop from unfertilized haploid eggs laid by queens or workers. In contrast, some drones develop from diploid eggs formed by the fusion of the ovum with one of the polar bodies or from fertilized eggs that are homologous at the sex locus (Woyke *et al.*, 1966; Herrmann *et al.*, 2005). Diploid drones can have uniparental origins or biparental origins. Further, the bi-parental origin diploid drone can create by matchmaking a queen and drones with identical sex alleles from the same colony.

Generally, brood-attending worker honey bees eliminate the false diploid drones (Woyke, 1962; Woyke, 1965; Woyke, 1963 a, b, c, d; Herrmann *et al.*, 2005). Additionally, the diploid drones produce more cuticular hydrocarbons than the workers (Santomauro *et al.*, 2004). The diploid drones produce diploid spermatozoa, having twice the DNA and an elongated head.

Fig. (1). Hexagonal Wax Cells, Worker Honey Bees, Drone Honey Bees and Ripe Honey Cells are depicted in this image. Worker honey bees perform different duties like the exchange of information, honey processing, and adding worker jelly to developing worker larvae.

The diploid drone larvae secrete certain substances known as cannibalization substances that act as the highlighter of diploid drones that attract other workers for cannibalism (Woyke, 1967; Dietz, 1975; Dietz and Lovins, 1975; Bienefeld *et al.*, 1994, 2000; Santomauro *et al.*, 2004). According to Woyke (1969a,b), diploid drones can be reared outside a colony, re-introduced into the colony, and accepted by the colonial residents.

Hymenopteran drones are either haploid or diploid, with a meiotic gametogenesis; therefore, the drones contain a carbon copy of the maternal genomic content (Woyke and Skowronek, 1974). The spermatozoa of the diploid drones are diploid with double nuclear content (Woyke, 1973, 1974; 1975; Fahrenhorst, 1977; Trenczek *et al.*, 1989; Engels *et al.*, 1990; Piulachs *et al.*, 2003; Santomauro *et al.*, 2004; Herrmann *et al.* 2005).

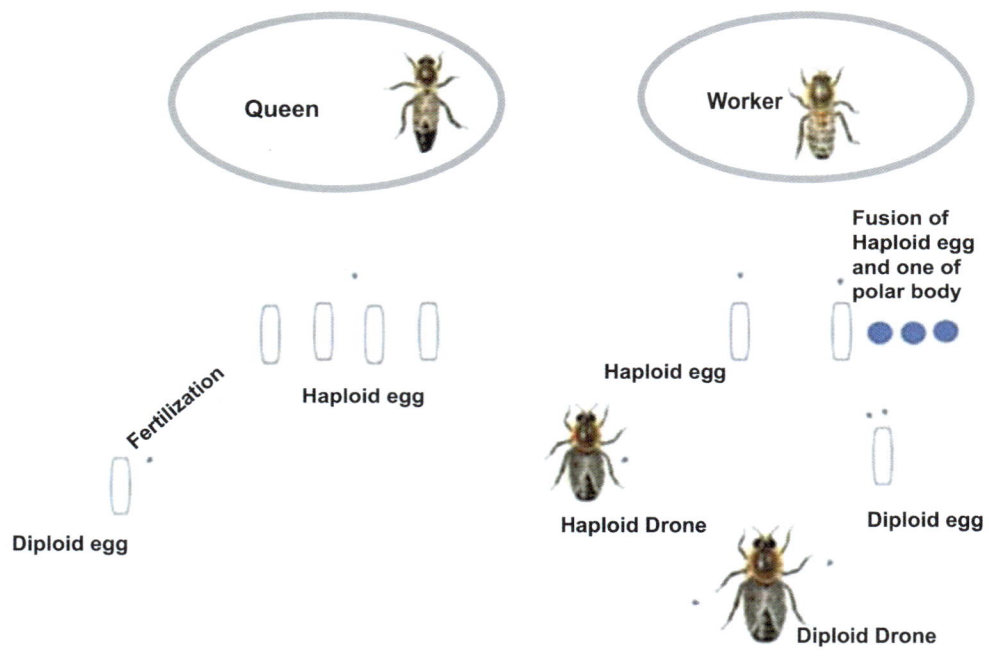

Fig. (2). Parthenogenesis and the sexual reproduction method in honey bees result in haploid and diploid drones forming.

LIFE EXPECTANCY IN DRONE HONEY BEE

The life span of drone honey is negatively proportional to extreme colonial and external environmental challenges. For example, the life expectancy of the drone honey bee varies from 13-14 days to 21-24 days (Howell and Usinger, 1933; Lavrek, 1947; Kepena,1963; Witherell, 1965; Drescher, 1969; Fukuda and

Ohtani, 1997; Herrmann *et al.*, 2005). Additionally, other factors that affect the life span of drones include flight activity or geographical region (Fukuda and Ohtani, 1977). For instance, the life span of a drone is comparatively shorter in summer than in autumn due to constraints imposed by harsh summer conditions (Fukuda and Ohtani, 1977). Further, the drone's life span depends upon flight performance and energy consumed during the flight (Neukirch (1982).

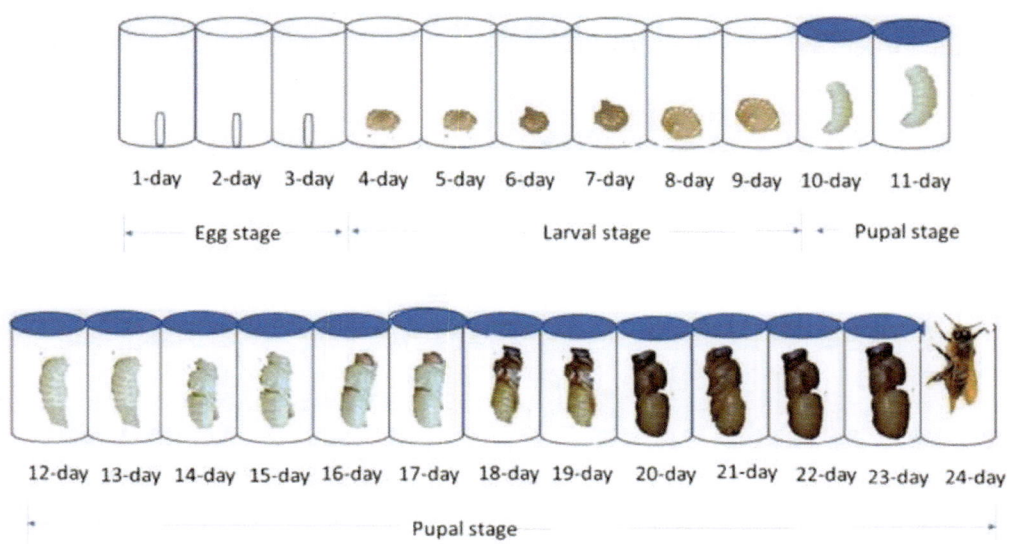

Fig. (3). Development synchronicity of drones honey bee in pre-capped and capped cells within 24 days. The drone honey bee development comprises egg, larval, pupal, and adult phases. Further developmental duration varies with the temperature of the hive.

DRONE POPULATION IN HONEY BEE COLONY

The construction of drone cells or their conversion to worker cells depends on age, fecundity, and the queen's mandibular gland pheromone (Darchen, 1960; Chauvin, 1961; Free, 1967). However, the construction of a drone comb is limited by the colony's strength and the number of drone cells already present in the hive (Allen, 1958; Free and Williams, 1975). Furthermore, worker honey bees control the number of drone eggs, larvae, and pupae (Free and Williams, 1975; Fukuda and Ohtani, 1977). The survival of drone cells in queen-less and queen-right colonies is the same in spring and summer, but in autumn, the drone brood survival rate is higher in queen-less colonies than in queen-right colonies (Wovke *et al.*, 1965a). Drone brood production and emergence depend upon the colony's queen, temperature, pollen grain and nectar storage (Gorbaczaw, 1961; Taber, 1964; Louveux *et al.*, 1973; Mesqum, 1976; Fukuda and Ohtani, 1977).

Therefore, for large drone brood production, colonies must have a high sugar syrup supply, adequate pollen availability, and a queenless condition. In a larger colony, on average, about 1500 adult drones are present(Currie,1982).

The workers regulate the population of drone honey bees in the colony by influencing rearing and evicting drones from the colony during a scarcity of nectar, disease, pest infestation or under other environmental stress challenges. During the eviction, honey bees first push the drones honey bees toward the periphery, then to walls, bottom boards, and finally from the colony (Levenets, 1956).

Sometimes worker's honey bees aggressively expel the drones from the colony by chewing, mauling, and pulling them out of the colony (Morse *et al.*, 1967). Generally, expulsion occurs slowly, taking several weeks in the autumn. A typical colony generally evict about 10-15 drones daily (MorsE *et al.*, 1967; Free and Williams, 1975).

Generally, temperature, nurse honey bee population, queen age, sealed and unsealed brood drone odour colony activity, available food in the colony, available honey, and genetic strains of bees are all factors that influence the eviction process (Levenets, 1951; Alber, 1955; Orosi, 1959; Moore *et al.*, 1967; Free and Williams, 1975; Free, *et al.*, 1977; Taber, 1982).

FLIGHT ACTIVITY OF DRONE HONEY BEES

The drone honey bees begin their flight activity at the age of four days, but the first proper flight usually occurs when drones are 5-7 days old (Howell and Usinger, 1933; Laverekhin, 1947; Taber, 1964; Drescher *et al.*, 1969; Witherell, 1972). Generally, drones' flight time varies in different species like *Apis mellifera, Apis cerana, Apis florea,* and *Apis dorsata* (Lavrekhin, 1947; Ruttner *et al.*, 1972; Koeniger and Sekera, 1976).

In *Apis mellifera,* European honey bee drones flights begin between 11.00–14.00 hr and usually end between 16.00–18.00 hr (Kurennoi, 1953; Oertel, 1956; Ruttner, 1966; Drescher, 1969; Strang, 1970; Callow, 1964; Avitable and Kasinskas, 1977). The time of flight is influenced by other environmental conditions, time of the day and the direction of the hive entrance. Additionally, drone flight is influenced by daily temperature, relative humidity, the colony's position along the sun's direction, and the day's length. Additionally, circadian rhythm influences the drone's flight activity. Finally, the drone honey bee's flight duration varies with the drone's age, as the oldest drones take the longest flights.

DRIFT AND ORIENTATION OF DRONES' FLIGHT TO THE HIVE

The drone honey bee can fly up to 7 km from its native colony and successfully return to the hive. The drone bees can use landmarks, the sun, or the earth's magnetic compass to find a hive. However, generally, drones make orientation errors and usually enter into incorrect colonies, known as drifting. Drifting drones can affect honey productivity as drones can act as a vector of many diseases and parasites, including sacbrood (virus), Nosema, Acarapis, and Varroa when they drift between hives and between apiaries.

In apiaries with hives placed in straight rows usually provide no orientation cues and can induce drone drifting. However, compared to a straight row of hives, the horseshoe layout arrangement of the hive reduces drone drift by only 10%.

Usually, the drone drifting process begins on 5-7-day old drones with the beginning of flight activity. In older drones, there is less drifting with an increase in age, which could be due to more flight experience. Younger drones make the highest flight errors compared to older ones. Generally, drone drift is less observable in areas with more orientation cues.

Drones tend to drift south with east and west entrances, whereas drones tend to drift west with entrances facing north and south. The directions of drones drifting appear to be influenced by the sun's position in the sky. The pheromone 9-ox--trans-2-decenoic acid influences drone drift.

Fig. (4). The hive section shows uncapped larval instars of honey bees and worker honey bees perfor-ming various tasks.

Fig. (5). The image illustrating different instar larvae and unripe honey cells are highlighted in a hive section.

REPRODUCTIVE SYSTEM OF THE DRONE HONEY BEES

Drone honey bee produces about 10 million male sperm cells that are genetically identical. The sperm cells are genetically similar to full sisters in the hive, which carries maternal and paternal inheritance. The drone honey bees solicit food from the workers after hatching (Figs. **1-5**). The drones honey bees feed during day before the flight, mainly on honey (Fukuda and Ohtani, 1977).

The male reproductive system of the drone consists of a pair of testes, a pair of vasa differentia, a pair of accessory sex glands, and a median ejaculatory duct (Bishop, 1920; Snodgrass, 1956; Woyke, 1958; Simpson, 1960; Chen, 1984; Davey, 1985; Koeniger, 1986; Gillot, 1988; Koeniger *et al.*, 1989; Paliwal, 1993; Chapman, 1998).

Testes

The gonads are creamy, oval-shaped bodies lying on the mucus glands' and are situated between the 2nd and 3rd abdominal segments. The number of testicular follicles is seven in *Apis cerana indica* and is variable in number in other hymenopterans (Snodgrass, 1956; Wheeler and Krutzsch, 1992; Duchateau and Mariën, 1995; Ferreira *et al.*, 2004). The testes contain seven tube-like coiled follicles.

Vas Deferens

It is divided into three regions: the apical short coiled tube, middle cylindrical seminal vesicle (SV), and distal straight duct. The apical part of VD possesses an outer circular muscle layer, an inner epithelial layer, and a large lumen.

Seminal Vesicle

It is represented by a sizeable sac-like region of the vas deferens. The seminal vesicle comprises inner epithelial layers lined by outer muscle layers and externally covered with a thin peritoneal sheath. Further, the epithelial layer comprises tall columnar cells with a luminal surface and a lumen filled with sperm bundles. The muscle layers are composed of outer longitudinal and inner circular muscle layers. The epithelial cells are tall and columnar with a brush border towards the lumen. The lumen is filled with a large mass of sperm bundles, with their heads towards the wall and tails at the centre of the lumen.

Mucous Glands

A pair of mucous glands (MG) that are kidney-shaped, sac-like structures representing a particular type of male accessory gland in the bee. The wall of the MG is composed of an inner epithelial layer and an outer thick muscle coat and is externally covered with a thin peritoneal sheath. The muscle coat comprises three sublayers; the outer longitudinal, middle circular, and inner longitudinal muscle layer (Bishop, 1920; Snodgrass, 1956; Paliwal, 1993; Paliwal, 1993; Sawarkar and Tembhare, 2015).

Ejaculatory Duct

The ejaculatory duct is a long, slender tube, with the wall of the ejaculatory duct differentiated into an outer broad epithelial layer and an inner cuticular layer. The inner thin cuticular layer bears elongated spines in the lumen. A cuticular layer showed that the ejaculatory duct is ectodermal in origin.

MATING

In honey bees, queen is polyandrous that mate with multiple drones during nuptial flight. For mating flight, the queen flies about 3 km from her hive to the congregation areas that could be approximately 5–40 m above the group (Currie, 1989; Loper *et al.*, 1992; Ellis *et al.*, 2015). Various explorations highlight variation in drone mating number. According to reports, the queen mates with 12 drones, whereas few reports indicate that the queen mates with 34-77 drones (Woyke and Ruttner, 1958; Winston, 1987; Kraus *et al.*, 2005; Ellis *et al.*, 2015; Withrow *et al.*, 2018).

Woyke (1962), Winston (1987), Schluns *et al.* (2005), and Ellis *et al.* (2015) estimate that 3-5 per cent of sperm from a single drone is stored for fertilization. Queen can hold about 4-7 million total sperms in her spermatheca (Koeniger, and Koeniger, 2000; Rhodes, 2002;Baer, 2005; Richard *et al.*, 2007; Ellis *et al.*, 2015). The honey bee colonies headed by multiple drones inseminated queens show more robust retinue response, possess more vital wax cell construction ability, produce more honey, reproduce fast, collect more pollen grain, produce more drones, and have more resistance to diseases in comparison to the colony headed by a queen inseminated by a single drone (Tarpy., 2003; Seeley, and Tarpy, 2006;Richard *et al.*, 2007; Seeley, and Tarpy, 2007; Schmid *et al.*, 2008; Nino *et al.*, 2012; Collins and Pettis, 2013; Cornman *et al.*, 2013; Niño *et al.*, 2013; Brutscher *et al.*, 2015; Mattila, and Seeley,2015; Kurze *et al.*, 2016; Peng *et al.*, 2016; Kurze *et al.*, 2016; Rueppell *et al.*, 2016; Brutscher *et al.*, 2017; McMenamin *et al.*, 2018; Brutscher *et al.*, 2019) .

DRONE AGING

The drone honey bee's reproductive quality is affected by age, season and genetics. Drone ageing has been shown to reduce sperm viscosity, volume, and viability (Woyke and Jasiski 1978; Locke and Peng 1993; Rhodes, 2002; Cobey 2007; Rhodes *et al.* 2011; Czeko ska *et al.* 2013; Stürup *et al.* 2013). In more aged drones, semen becomes dark and more viscous than in younger drones (Woyke and Jasiski 1978; Cobey 2007; Czekoska *et al.* 2013a). This type of sperm forms plugs in the oviduct and interferes with reproduction (Woyke and Jasiski 1978; Czekoska *et al.* 2013a; Stürup *et al.* 2013; Metzand Tarpy, 2019). Locke and Peng (1993) detected that ageing affects sperm viability, decreasing from 86% to 81% in 14-day-old and 20-day-old drones, respectively (Woyke and Jasiski 1978; Locke and Peng 1993; Cobey 2007; Rhodes *et al.* 2011; Czekoska *et al.* 2013a; Metzand Tarpy 2019; Rangel 2019).

CONCLUSION

Haploid drone honey bees develop from unfertilized eggs through parthenogenesis and diploid drones from diploid eggs if the queen mates with drones of the same colony or from diploid eggs of workers. The unfertilized eggs for drone development can be laid by a queen, pseudo queen or worker honey bees. During development, drone larvae are initially fed on royal jelly, and after that, honey, pollen grains, and drone jelly with a composition different from royal jelly. Drones perform the duty of mating with queen honey bees when they get attracted to her, and further, drones cooperate in temperature regulation of the colony.

REFERENCES

Adams, J, Rothman, ED, Kerr, WE & Paulino, ZL (1977) Estimation of the number of sex alleles and queen matings from diploid male frequencies in a population of Apis mellifera. *Genetics*, 86, 583-96.
[http://dx.doi.org/10.1093/genetics/86.3.583] [PMID: 892423]

Akyol, E, Yeninar, H & Kaftanoglu, O (2008) Live weight of queen honey bees (Apis mellifera L.) predicts reproductive characteristics. *J Kans Entomol Soc*, 81, 92-100.
[http://dx.doi.org/10.2317/JKES-705.13.1]

Alber, MA (1955) *Südwestdeurscher Imker*, 7, 106-7. [Drone colonies]. [In German.].

Al-Lawati, H, Kamp, G & Bienefeld, K (2009) Characteristics of the spermathecal contents of old and young honeybee queens. *J Insect Physiol*, 55, 117-22.
[http://dx.doi.org/10.1016/j.jinsphys.2008.10.010] [PMID: 19027748]

Allen, DM (1958) Drone brood in honey bee colonies. *J Econ Entomol*, 51, 46-8.
[http://dx.doi.org/10.1093/jee/51.1.46]

Al-Qarni, AS, Phelan, PL, Smith, BH & Cobey, SW (2005) The influence of mating type and oviposition period on mandibular pheromone levels in Apis mellifera L. honeybee queens. *Saudi J Biol Sci*, 12, 39-47.

Amiri, E, Meixner, MD & Kryger, P (2016) Deformed wing virus can be transmitted during natural mating in honey bees and infect the queens. *Sci Rep*, 6, 33065.
[http://dx.doi.org/10.1038/srep33065] [PMID: 27608961]

Avitabile, A & Kasinskas, JR (1977) The drone population of natural honeybee swarms. *J Apic Res*, 16, 145-9.
[http://dx.doi.org/10.1080/00218839.1977.11099876]

Baer, B (2005) Sexual selection in *Apis* bees. *Apidologie (Celle)*, 36, 187-200.
[http://dx.doi.org/10.1051/apido:2005013]

Bishop, GH (1920) Fertilization in the honey-bee. I. The male sexual organs: Their histological structure and physiological functioning. *J Exp Zool*, 31, 224-65.
[http://dx.doi.org/10.1002/jez.1400310203]

Brutscher, LM, Baer, B & Niño, EL (2019) Putative drone copulation factors regulating honey bee (Apis mellifera) queen reproduction and health: A review. *Insects*, 10, 8.
[http://dx.doi.org/10.3390/insects10010008] [PMID: 30626022]

Brutscher, LM, Daughenbaugh, KF & Flenniken, ML (2015) Antiviral defense mechanisms in honey bees. *Curr Opin Insect Sci*, 10, 71-82.
[http://dx.doi.org/10.1016/j.cois.2015.04.016] [PMID: 26273564]

Brutscher, LM, Daughenbaugh, KF & Flenniken, ML (2017) Virus and dsRNA-triggered transcriptional responses reveal key components of honey bee antiviral defense. *Sci Rep*, 7, 6448.
[http://dx.doi.org/10.1038/s41598-017-06623-z] [PMID: 28127051]

Callow, RK, Chapman, JR & Paton, AN (1964) Pheromones of the honeybee: chemical studies of the mandibular gland secretion of the queen. *J Apic Res*, 3, 77-89.
[http://dx.doi.org/10.1080/00218839.1964.11100086]

Chaud-Netto, J (1975) Sex determination in bees. II. Additivity of maleness genes in Apis mellifera. *Genetics*, 79, 213-7.
[http://dx.doi.org/10.1093/genetics/79.2.213] [PMID: 1132678]

Collins, AM & Pettis, JS (2013) Correlation of queen size and spermathecal contents and effects of miticide exposure during development. *Apidologie (Celle)*, 44, 351-6.
[http://dx.doi.org/10.1007/s13592-012-0186-1]

Cornman, RS, Lopez, D & Evans, JD (2013) Transcriptional response of honey bee larvae infected with the bacterial pathogen Paenibacillus larvae. *PLoS One*, 8, e65424.

[http://dx.doi.org/10.1371/journal.pone.0065424] [PMID: 23762370]

Currie, RW *Some factors affecting the orientation of drone honey bees (Apis mellifera L.)* (Master's thesis).

Darchen, R (1957) The queen of Apis mellificates egg-laying workers and wax constructions. *Insectes Soc,* 4, 321-5.
[http://dx.doi.org/10.1007/BF02224152]

de Miranda, JR & Fries, I (2008) Venereal and vertical transmission of deformed wing virus in honeybees (Apis mellifera L.). *J Invertebr Pathol,* 98, 184-9.
[http://dx.doi.org/10.1016/j.jip.2008.02.004] [PMID: 18358488]

Drescher, W (1969) The flight activity of Apis melifera carnica and Apis mellifera ligustica drones in relation to age and weather. *Z Bienenf,* 9, 390-409.

Ellis, J, Lawrence, JC, Koeniger, N & Koeniger, G Mating Biology of Honey Bees (Apis mellifera). *American Bee Journal.*

Free, JB & Williams, IH (1975) Factors determining the rearing and rejection of drones by the honeybee colony. *Anim Behav,* 23, 650-75.
[http://dx.doi.org/10.1016/0003-3472(75)90143-8]

Free, JB (1967) The production of drone comb by honeybee colonies. *J Apic Res,* 6, 29-36.
[http://dx.doi.org/10.1080/00218839.1967.11100157]

Fukuda, H & Ohtani, T (1977) Survival and life span of drone honeybees. *Popul Ecol,* 19, 51-68.
[http://dx.doi.org/10.1007/BF02510939]

Garofalo, C (1972) Behaviour and sexual maturity of Apis mellifera adansonii drones.*Homenagem.*Faculty of Philosophy, Sciences and Letters, Rio Claro, Brazil.

Gilley, DC, Tarpy, DR & Land, BB (2003) Effect of queen quality on interactions between workers and dueling queens in honeybee (Apis mellifera L.) colonies. *Behav Ecol Sociobiol,* 55, 190-6.
[http://dx.doi.org/10.1007/s00265-003-0708-y]

(1961) Colonies without drones—future beekeeping. *Pszczelarstwo,* 12, 19-21.

Gregorc, A & Škerl, MI (2015) Characteristics of honey bee (Apis mellifera Carnica, Pollman 1879) queens reared in Slovenian commercial breeding stations. *J Apic Sci,* 59, 5-12.

Haarmann, T, Spivak, M, Weaver, D, Weaver, B & Glenn, T (2002) Effects of fluvalinate and coumaphos on queen honey bees (Hymenoptera: Apidae) in two commercial queen rearing operations. *J Econ Entomol,* 95, 28-35.
[http://dx.doi.org/10.1603/0022-0493-95.1.28] [PMID: 11942761]

Hatch, S, Tarpy, DR & Fletcher, DJC (1999) Worker regulation of emergency queen rearing in honey bee colonies and the resultant variation in queen quality. *Insectes Soc,* 46, 372-7.
[http://dx.doi.org/10.1007/s000400050159]

Herrmann, M, Trenzcek, T, Fahrenhorst, H & Engels, W (2005) Characters that differ between diploid and haploid honey bee (*Apis mellifera*) drones. *Genet Mol Res,* 4, 624-41.
[PMID: 16475107]

Howell, DE & Usinger, RL (1933) Observations on the flight and length of life of drone bees. *Ann Entomol Soc Am,* 26, 239-46.
[http://dx.doi.org/10.1093/aesa/26.2.239]

Jay, SC (1963) The development of honeybees in their cells. *J Apic Res,* 2, 117-34.
[http://dx.doi.org/10.1080/00218839.1963.11100072]

Johnson, JN, Hardgrave, E, Gill, C & Moore, D (2010) Absence of consistent diel rhythmicity in mated honey bee queen behavior. *J Insect Physiol,* 56, 761-73.
[http://dx.doi.org/10.1016/j.jinsphys.2010.01.004] [PMID: 20116381]

Kahya, Y, Gençer, HV & Woyke, J (2008) Weight at emergence of honey bee (*Apis mellifera caucasica*) queens and its effect on live weights at the pre and post mating periods. *J Apic Res,* 47, 118-25. [http://dx.doi.org/10.1080/00218839.2008.11101437]

Katzav-Gozansky, T, Soroker, V & Hefetz, A (1997) The biosynthesis of Dufour's gland constituents in queens of the honeybee (Apis mellifera). *Invertebrate Neuroscience,* 3, 239-43.

Keeling, CI, Slessor, KN, Higo, HA & Winston, ML (2003) New components of the honey bee (*Apis mellifera* L.) queen retinue pheromone. *Proc Natl Acad Sci USA,* 100, 4486-91.
[http://dx.doi.org/10.1073/pnas.0836984100] [PMID: 12676987]

Kepena, L (1963) On the biology of drones of the Tatranka breed. *Proceedings XIX International Beekeeping Congress,* 307-8.

Kerr, WE (1974) Advances in cytology and genetics of bees. *Annu Rev Entomol,* 19, 253-68.
[http://dx.doi.org/10.1146/annurev.en.19.010174.001345]

Kerr, WE & Silveira, ZV (1974) A note on the formation of honeybee spermatozoa. *J Apic Res,* 13, 121-6.
[http://dx.doi.org/10.1080/00218839.1974.11099767]

Kocher, SD, Richard, FJ, Tarpy, DR & Grozinger, CM (2009) Queen reproductive state modulates pheromone production and queen-worker interactions in honeybees. *Behav Ecol,* 20, 1007-14.
[http://dx.doi.org/10.1093/beheco/arp090] [PMID: 22476212]

Kocher, SD, Richard, FJ, Tarpy, DR & Grozinger, CM (2008) Genomic analysis of post-mating changes in the honey bee queen (Apis mellifera). *BMC Genomics,* 9, 232.
[http://dx.doi.org/10.1186/1471-2164-9-232] [PMID: 18489784]

Koeniger, N (1969) Experiments concerning the ability of the queen (*Apis mellifera* L.) to distinguish between the drone and worker cells. *XXII International Beekeeping Congress Summit,* 138.

Koeniger, N & Koeniger, G (2000) Reproductive isolation among species of the genus *Apis. Apidologie,* 31, 313-39.
[http://dx.doi.org/10.1051/apido:2000125]

Kurennoj, NM [Flight activity and sexual maturity of drones.) Pche/ovodstvo 31pp. 24-28

Kurze, C, Dosselli, R, Grassl, J, Le Conte, Y, Kryger, P & Baer, B (2016) & Moritz RF. Differential proteomics reveals novel insights into Nosema–honey bee interactions. *Insect Biochem Mol Biol* 42-9.

Lavrekm, FA (1960) A problem of the biology of drones. *Pchelovodstvo,* 24, 57-9.

Lavrekm, FA (1960) Comparative observations on the flight activity of drones. Pche/ovodstvo. 37, 43-5.

Levenets, IP (1956) *Pchelovodstvo,* 33, 28-9. [Observations on the expulsion of drones.]. [In Russian.].

Lodesani, M, Balduzzi, D & Galli, A (2004) Functional characterisation of semen in honeybee queen (*A.m.ligustica* S.) spermatheca and efficiency of the diluted semen technique in instrumental insemination. *Ital J Anim Sci,* 3, 385-92.
[http://dx.doi.org/10.4081/ijas.2004.385]

Louveaux, J, Mesquida, J & Fresnaye, J (1972) Observations on the variability of male brood production in bee colonies (Apis mellifica L.). *Apidologie,* 3, 291-307.
[http://dx.doi.org/10.1051/apido:19720401]

Mackensen, O (1964) Relation of semen volume to success in artificial insemination of queenhoney bees. *J Econ Entomol,* 571, 581-3.

Mazeed, AM & Mohanny, KM (2010) Some reproductive characteristics of honeybee drones in relation to their ages. *Entomol Res,* 40, 245-50.
[http://dx.doi.org/10.1111/j.1748-5967.2010.00297.x]

McMenamin, A, Daughenbaugh, K, Parekh, F, Pizzorno, M & Flenniken, M (2018) Honey bee and bumble bee antiviral defense. *Viruses,* 10, 395-407.
[http://dx.doi.org/10.3390/v10080395] [PMID: 30060518]

MesQum, AJ (1976) Further observations on drone rearing in honey bee colonies. *Apidologie,* 7, 307-30.

Metz, B & Tarpy, D (2019) Reproductive senescence in drones of the honey bee (Apis mellifera). *Insects,* 10, 11.
[http://dx.doi.org/10.3390/insects10010011] [PMID: 30626026]

Morse, RA, Strang, GE & Nowakowski, J (1967) Fall death rates of drone honeybees. *J Econ Entomol,* 60, 1198-202.
[http://dx.doi.org/10.1093/jee/60.5.1198]

Niño, EL, Malka, O, Hefetz, A, Teal, P, Hayes, J & Grozinger, CM (2012) Effects of honey bee (Apis mellifera L.) queen insemination volume on worker behavior and physiology. *J Insect Physiol,* 58, 1082-9.
[http://dx.doi.org/10.1016/j.jinsphys.2012.04.015] [PMID: 22579504]

Niño, EL, Malka, O, Hefetz, A, Tarpy, DR & Grozinger, CM (2013) Chemical profiles of two pheromone glands are differentially regulated by distinct mating factors in honey bee queens (*Apis mellifera* L.). *PLoS One,* 8, e78637.
[http://dx.doi.org/10.1371/journal.pone.0078637] [PMID: 24236028]

Niño, EL, Tarpy, DR & Grozinger, CM (2013) Differential effects of insemination volume and substance on reproductive changes in honey bee queens (*Apis mellifera* L.). *Insect Mol Biol,* 22, 233-44.
[http://dx.doi.org/10.1111/imb.12016] [PMID: 23414204]

Oertel, E (1956) Observations on the flight of drone honey bees. *Ann Entomol Soc Am,* 49, 497-500.
[http://dx.doi.org/10.1093/aesa/49.5.497]

Pál, ZÖ (1959) The behaviour and nutrition of drones. *Bee World,* 40, 141-6.
[http://dx.doi.org/10.1080/0005772X.1959.11096717]

Page, RE, Jr & Laidlaw, HH, Jr (1985) Closed population honeybee breeding. *Bee World,* 66, 63-72.
[http://dx.doi.org/10.1080/0005772X.1985.11098826]

Page, RE, Jr & Peng, CYS (2001) Aging and development in social insects with emphasis on the honey bee, Apis mellifera L. *Exp Gerontol,* 36, 695-711.
[http://dx.doi.org/10.1016/S0531-5565(00)00236-9] [PMID: 11295509]

Pankiw, T, Winston, ML, Plettner, E, Slessor, KN, Pettis, JS & Taylor, OR (1996) Mandibular gland components of european and africanized honey bee queens (Apis mellifera L.). *J Chem Ecol,* 22, 605-15.
[http://dx.doi.org/10.1007/BF02033573] [PMID: 24227572]

Peer, DF (1956) Multiple matings of queen honeybees. *J Econ Entomol,* 49, 741-3.
[http://dx.doi.org/10.1093/jee/49.6.741]

Peng, Y, Grassl, J & Millar, AH (2015) Seminal fluid of honey bees contains multiple mechanisms to combat infections of the sexually transmitted pathogen Nosema apis. *Proc Biol Sci,* 283, 1785.

Rangel, J, Böröczky, K, Schal, C & Tarpy, DR (2016) Honey bee (*Apis mellifera*) queen reproductive potential affects queen mandibular gland pheromone composition and worker retinue response. *PLoS One,* 11, e0156027.
[http://dx.doi.org/10.1371/journal.pone.0156027] [PMID: 27281328]

Renner, M & Baumann, M (1964) About complexes of subepidermal gland cells (scent glands?) of the queen bee. *Natur wissen schaften,* 51, 68-9.
[http://dx.doi.org/10.1007/BF00603470]

Rhodes, J (2002) *Drone Honey Bees-Rearing and Maintenance.*NSW Agriculture, Orange, Australia.

Richard, FJ, Schal, C, Tarpy, DR & Grozinger, CM (2011) Effects of instrumental insemination and insemination quantity on Dufour's gland chemical profiles and vitellogenin expression in honey bee queens (*Apis mellifera*). *J Chem Ecol,* 37, 1027-36.
[http://dx.doi.org/10.1007/s10886-011-9999-z] [PMID: 21786084]

Richard, FJ, Tarpy, DR & Grozinger, CM (2007) Effects of insemination quantity on honey bee queen

physiology. *PLoS One,* 2, e980.
[http://dx.doi.org/10.1371/journal.pone.0000980] [PMID: 17912357]

Roberts, KE, Evison, SEF, Baer, B & Hughes, WOH (2015) The cost of promiscuity: sexual transmission of Nosema microsporidian parasites in polyandrous honey bees. *Sci Rep,* 5, 10982.
[http://dx.doi.org/10.1038/srep10982] [PMID: 26123939]

Rueppell, O, Aumer, D & Moritz, RFA (2016) Ties between ageing plasticity and reproductive physiology in honey bees (*Apis mellifera*) reveal a positive relation between fecundity and longevity as consequence of advanced social evolution. *Curr Opin Insect Sci,* 16, 64-8.
[http://dx.doi.org/10.1016/j.cois.2016.05.009] [PMID: 27720052]

Ruttner, F, Woyke, J & Koeniger, N (1972) Reproduction in Apis cerana 1. Mating behaviour. *J Apic Res,* 11, 141-6.
[http://dx.doi.org/10.1080/00218839.1972.11099714]

Ruttner, F (1966) The life and flight activity of drones. *Bee World,* 47, 93-100.
[http://dx.doi.org/10.1080/0005772X.1966.11097111]

Sandrock, C, Tanadini, M, Tanadini, LG, Fauser-Misslin, A, Potts, SG & Neumann, P (2014) Impact of chronic neonicotinoid exposure on honeybee colony performance and queen supersedure. *PLoS One,* 9, e103592.
[http://dx.doi.org/10.1371/journal.pone.0103592] [PMID: 25084279]

Santomauro, G, Oldham, NJ, Boland, W & Engels, W (2004) Cannibalism of diploid drone larvae in the honey bee (*Apis mellifera*) is released by odd pattern of cuticular substances. *J Apic Res,* 43, 69-74.
[http://dx.doi.org/10.1080/00218839.2004.11101114]

Sawarkar, AB & Tembhare, DB (2015) Histomorphological study of the male reproductive system in the Indian drone Honeybee, Apis cerana indica (Hymenoptera). *National Conference on Advances in Bioscience & Environmental Science: Present & Future (ABES),* 63.

Schlüns, H, Moritz, RFA, Neumann, P, Kryger, P & Koeniger, G (2005) Multiple nuptial flights, sperm transfer and the evolution of extreme polyandry in honeybee queens. *Anim Behav,* 70, 125-31.
[http://dx.doi.org/10.1016/j.anbehav.2004.11.005]

Schmid, MR, Brockmann, A, Pirk, CWW, Stanley, DW & Tautz, J (2008) Adult honeybees (Apis mellifera L.) abandon hemocytic, but not phenoloxidase-based immunity. *J Insect Physiol,* 54, 439-44.
[http://dx.doi.org/10.1016/j.jinsphys.2007.11.002] [PMID: 18164310]

Seeley, TD & Tarpy, DR (2007) Queen promiscuity lowers disease within honeybee colonies. *Proc Biol Sci,* 274, 67-72.
[http://dx.doi.org/10.1098/rspb.2006.3702] [PMID: 17015336]

Slessor, KN, Kaminski, LA, King, GGS & Winston, ML (1990) Semiochemicals of the honeybee queen mandibular glands. *J Chem Ecol,* 16, 851-60.
[http://dx.doi.org/10.1007/BF01016495] [PMID: 24263600]

Strang, GE (1970) A study of honey bee drone attraction in the mating response. *J Econ Entomol,* 63, 641-5.
[http://dx.doi.org/10.1093/jee/63.2.641]

Taber, S, III (1964) Factors influencing the circadian flight rhythm of drone honey bees. *Ann Entomol Soc Am,* 57, 769-75.
[http://dx.doi.org/10.1093/aesa/57.6.769]

Tanaka, ED & Hartfelder, K (2004) The initial stages of oogenesis and their relation to differential fertility in the honey bee (*Apis mellifera*) castes. *Arthropod Struct Dev,* 33, 431-42.
[http://dx.doi.org/10.1016/j.asd.2004.06.006] [PMID: 18089049]

Tarpy, DR (2003) Genetic diversity within honeybee colonies prevents severe infections and promotes colony growth. *Proc Biol Sci,* 270, 99-103.
[http://dx.doi.org/10.1098/rspb.2002.2199] [PMID: 12596763]

Tarpy, DR, Hatch, S & Fletcher, DJC (2000) The influence of queen age and quality during queen replacement in honeybee colonies. *Anim Behav,* 59, 97-101.
[http://dx.doi.org/10.1006/anbe.1999.1311] [PMID: 10640371]

Tarpy, DR & Seeley, TD (2006) Lower disease infections in honeybee (*Apis mellifera*) colonies headed by polyandrous vs monandrous queens. *Naturwissenschaften,* 93, 195-9.
[http://dx.doi.org/10.1007/s00114-006-0091-4] [PMID: 16518641]

de Oliveira Tozetto, S, Bitondi, MMG, Dallacqua, RP & Simões, ZLP (2007) Protein profiles of testes, seminal vesicles and accessory glands of honey bee pupae and their relation to the ecdysteroid titer. *Apidologie (Celle),* 38, 1-11.
[http://dx.doi.org/10.1051/apido:2006045]

Winston, ML (1987) *The Biology of the Honey Bee.*Harvard University Press, Cambridge, MA, USA.

Witherell, PC (1965) Survival of drones following eversion. *Ann Abeille,* 8, 317-20.
[http://dx.doi.org/10.1051/apido:19650405]

Witherell, P C (1972) Flight activity and natural monality of normal and mutant drone honeybees. *J Apic Res,* 11, 65-75.

Withrow, JM & Tarpy, DR (2018) Cryptic "royal" subfamilies in honey bee (*Apis mellifera*) colonies. *PLoS One,* 13, e0199124.
[http://dx.doi.org/10.1371/journal.pone.0199124] [PMID: 29995879]

Woyke, J (1962) Natural and Artificial Insemination of Queen Honeybees. *Bee World,* 43, 21-5.
[http://dx.doi.org/10.1080/0005772X.1962.11096922]

Woyke, J (1962) The hatchability of "lethal" eggs in a two-sex allele fraternity of honeybees. *J Apic Res,* 1, 6-13. a
[http://dx.doi.org/10.1080/00218839.1962.11100040]

Woyke, J (1963) Rearing and viability of diploid drone larvae. *J Apic Res,* 2, 77-84. b
[http://dx.doi.org/10.1080/00218839.1963.11100064]

Woyke, J (1963) What happens to diploid drone larvae in a honeybee colony? *J Apic Res,* 2, 73-5. c
[http://dx.doi.org/10.1080/00218839.1963.11100063]

Woyke, J (1965) Study on the comparative viability of diploid and haploid larval drone honeybees. *J Apic Res,* 4, 12-6.
[http://dx.doi.org/10.1080/00218839.1965.11100096]

Woyke, J (1974) Genic balance, heterozygosity and inheritance of testis size in diploid drone honeybees. *J Apic Res,* 13, 77-91.
[http://dx.doi.org/10.1080/00218839.1974.11099763]

Woyke, J (1977) Cannibalism and brood-rearing efficiency in the honeybee. *J Apic Res,* 16, 84-94.
[http://dx.doi.org/10.1080/00218839.1977.11099866]

Woyke, J (1978) Comparative biometrical investigation on diploid drones of the honeybee. II. The thorax. *J Apic Res,* 17, 195-205. a
[http://dx.doi.org/10.1080/00218839.1978.11099927]

Woyke, J (1978) Comparative biometrical investigation on diploid drones of the honeybee. III. The abdomen and weight. *J Apic Res,* 17, 206-17. b
[http://dx.doi.org/10.1080/00218839.1978.11099928]

Woyke, J & Ruttner, F (1958) An Anatomical Study of the Mating Process in the Honeybee. *Bee World,* 39, 3-18.
[http://dx.doi.org/10.1080/0005772X.1958.11095028]

Woyke, J & Knytel, A (1966) The chromosome number as proof that drones can arise from fertilized eggs of the honeybee. *J Apic Res,* 5, 149-54.

[http://dx.doi.org/10.1080/00218839.1966.11100148]

Woyke, J, Knytel, A & Bergandy, K (1966) The presence of spermatozoa in eggs as proof that drones can develop from inseminated eggs of the honeybee. *J Apic Res,* 5, 71-8.
[http://dx.doi.org/10.1080/00218839.1966.11100137]

Wu-Smart, J & Spivak, M (2016) Sub-lethal effects of dietary neonicotinoid insecticide exposure on honey bee queen fecundity and colony development. *Sci Rep,* 6, 32108.
[http://dx.doi.org/10.1038/srep32108] [PMID: 27562025]

Yue, C, Schröder, M, Bienefeld, K & Genersch, E (2006) Detection of viral sequences in semen of honeybees (Apis mellifera): Evidence for vertical transmission of viruses through drones. *J Invertebr Pathol,* 92, 105-8.
[http://dx.doi.org/10.1016/j.jip.2006.03.001] [PMID: 16630626]

CHAPTER 2

The Drone Honey Bee Morphometric Character

Abstract: The Drone caste exhibits specific diagnostic morphometric characteristics that facilitate its differentiation from other colony castes. Drone formation gets completed in about 24-25 days, with the first ten days in an open cell and the final 14 days under capped conditions. Furthermore, the drone caps are convex in shape. The Drone egg is measured about 1.49±0.12 (range 1.12–1.85) mm, the width is 0.35±0.02 (range 0.30–0.40) mm, and the volume is 0.10±0.02 (range 0.06–0.15) cubic mm. After hatching for the first three days, the larval weight is 0.11 mg, the 7th-day-old larva is 120 mg, 11th-day-old larva reaches a weight of 350 mg. An adult has a mean body length of about 1.5 cm. Further, the drone character varies as per ecological conditions, species, genotype, and other environmental conditions.

Keywords: Diploid drones, Haploid drones, Pheromones, Reproductive system.

INTRODUCTION

In a strong honey bee colony, usually with one queen, thousands of workers and a few hundred drones live harmoniously (Fig. **1**). The queen honey bee is polyandrous as she mates with multiple drones. When the queen of a colony is not reproductively active, workers start laying eggs and acting as pseudo-queens. Pheromones from the queen suppress the development of the workers' reproductive systems; therefore, workers cannot mate. As a result, workers can lay only unfertilized eggs, which can develop into drone honey bees (De Groot & Voogd, 1954; Butler, 1957; Jay, 1968; Pettis *et al.*, 1997). When a colony becomes queenless and workers cannot rear another queen, in that case, workers undergo ovarian development and become reproductively active to lay unfertilized haploid eggs, which results in drone production (Winston, 1987; Page & Erickson, 1988; Visscher, 1998).

In general, in honey bee colonies, two distinct-sized hexagonal cells are common, including small cells (5.2–5.4 mm in diameter) in which the queen lays fertilized eggs and large cells (6.2–6.4 mm in diameter) in which the queen lays unfertilized eggs (Winston, 1987). Even in queen-headed colonies, some workers lay unfertilized eggs in worker cells (Page & Erickson, 1988). Generally, such eggs are eaten by the other nestmate workers, and about 0.1% of drones are reared to adulthood by workers (Ratnieks & Visscher, 1989; Visscher, 1989). Typically, pseudo-queen formation occurs in queenless colonies, which begin laying unfertile eggs within 1-4 weeks, depending on the geographic region (Ruttner & Hesse, 1979). Hexagonal cell size determines the body size of emerging drones. Drones that have been reared in worker cells are smaller than those which emerge from other drone cells (Berg, 1991; Berg *et al.*, 1997; Schlüns *et al.*, 2003).

Drones are male honey bees without stings, and they do not perform hive duties. Structurally, drones are incapable of collecting food in the field. Food supply is stopped during late autumn when the colony experiences a dearth period. Drones are fed by worker honey bees in a colony. Due to food scarcity during winter, drones are not maintained in the colony. When drones become weak, their wings are torn off. Their legs are pulled and eventually dragged out of the hive. Drone larvae and pupae are removed from the hive sometimes. Their function is to mate with virgin queens on a mating flight.

The presence of healthy drones is a prime requirement of the colony for queen mating. The drone production depends upon the season and other environmental conditions of the colony and on the young and foraging workers' population. Furthermore, drone quality is influenced by divergent factors like species or subspecies of honey bees, age of drones, rearing season, food supply, size of comb cell, the infestation of comb, and colony strength. The queen honey bee mates with an average of 12 drones and can store 4.3-7.0 million spermatozoa in her spermatheca. Although the life span of a drone is about 55-90 days, and it gets sexual maturity at 16 days, it grows less suitable for mating. The prime role of the drone is to fertilize queens.

A ten-day-old drone is capable of impregnating a queen honey bee. Mating takes place in the air, and after mating, drones die. Copulation occurs about 12.8 kilometres from the hive in a drone congregational area. Drone produces different pheromones in drone congregational sizes. Therefore, producing viable drones is a limiting factor in successful queen rearing. Furthermore, the drone drifts to the colony when the colony possesses enough nectar, and pollen is permitted to stay.

Weight at maturity is considered a competitive advantage to drones over smaller males when fighting for access to females (BERG *et al.*, 1997). Therefore, Schluns *et al.* (2003) hypothesized that the low reproductive success of smaller drones is due to a low success rate in competition for accessing the queen.

Fig. (1). *Apis mellifera* strong colony with one exposed comb.

MORPHOMETRIC CHARACTERS

Haploid or diploid drones of honey bees possess certain specific morphometric features. The body is divided into a head, thorax, and abdomen, with three pairs of legs and two pairs of wings. Further explorations witnessed different morphometric features of drones. It can be easily speculated from these studies that drone body size varies as per ecological conditions. A few reports which deal with the morphometric analysis of drones are: Taha *et al.* (2012) studied *Apis mellifera jemenitica* (AMJ) colonies and drones' morphometric features of Carniolan and Yemeni. They reported that the body weight and cell size of Carniolan and Yemeni drones were 190.90 and 0.40 v.s 227.22 mg and 0.43 cm^3, respectively. Further, they speculated that the length and width of the right forewing and the number of hamuli on the right-hand wing of Carniolan drones were significantly higher than in Yemeni subspecies. Taha and Alqarni, 2013 compared the reproductive potential of Yemeni drones and Carniolan.

Furthermore, they reported that the length and width of the forewing of Yemeni drones were about 13.63% and 15.19% smaller than those of Carniolan ones. They noted the height and width of the hind wing and the number of hamuli of Yemeni drones were about 3.87%, 2.54% and 12.82%, smaller than those of Carniolan.

In addition to that, researchers also compared the reproductive features of Carniolan and Yemeni drones. According to them, the average size of Yemeni drone reproductive organs, including testis, seminal vesicle, and mucus gland, was 46.58, 55.81, and 34.53% smaller than Carniolan. Even the sperm number in the seminal vesicle of Yemeni drones has been reported to be 35.80% less than that of Carniolan ones.

Few studies witness the morphometric features of drones in three honey bee species, including *Apis dorsata F., Apis florea F., and Apis cerana*. For species determination, the structure of male genitalia is generally considered (Ruttner,1988). However, a few explorations related to the characteristic determination of male genitalia in *Apis florea, Apis dorsata and Apis cerana* are available in the scientific literature (Bahrmann (1961), Ruttner (1988), Simpson (1970), Koeniger and Koeniger (1990), Koeniger *et al.* (1991).

Ruttner (1983, 1988) and Viraktamath *et al.* (2015) studied the morphometrics of drones of *Apis mellifera, Apis florea, Apis dorsata* and *Apis cerana* for the identification of subspecies. Furthermore, Viraktamath *et al.* (2015) compared morphometric parameters and genitalia characteristics of drones from different states including Andhra Pradesh (AP), Assam (AS), Jammu & Kashmir (JK), Karnataka (KA), Kerala (KL), Maharashtra (MH) and Tamil Nadu (TN). They reported that in *Apis cerana* drones from Assam had the most extended body length at 11.47 mm, followed by Maharashtra at 10.82 mm and Jammu and Kashmir at 10.68 mm. Tamil Nadu drones have the shortest body length of 9.61 mm. The size and width of the drone from Assam were 0.85 and 3.86 mm, and those from Jammu and Kashmir, were 0.87 and 4.06 mm. Drones from Assam, Jammu, and Kashmir have longer femurs, hind tibias, and metatarsals than drones from other states. They further reported that the length and width of the metatarsus vary from 1.79 to 2.12 in Tamil Nadu, Jammu, and Kashmir.

Furthermore, morphometric analysis by Rutter, in 1988 indicated that the length of the hind leg of drones from South India corresponds to 7.56 mm. In contrast, Afghanistan, Pakistan, and North China drones possess an approximately 8.34 mm long hind limb. Viraktamath, 2015 reported that the forewings of drones from Jammu and Kashmir have longer forewings of 10.4 and 3.54 mm, respectively, whereas drones from Assam possess wings of lengths of 9.99 and 3.22 mm. The ratio of height to width of the forewing varied from 2.94 to 3.22.

Ruttner (1988) studied drones from Afghanistan, Pakistan, South India, and North China and concluded that drones from the North are more significantly larger than those from the South. Viraktamath, 2015 further concluded that in *Apis dorsata* the body length of drones varies in different states, as drones from Assam possess a minimum size of 12.45 mm and drones from Kerala have a maximum body length of 15.95 mm. Further, they concluded that the length of the vertex and width of the head varied from 1.08 to 1.17 mm and 4.52 to 4.68 mm in drones of different states. They determined that the distance between two lateral ocelli and lateral ocellus to eyes varies from 0.23 to 0.26 mm and 0.07 to 0.10 mm, respectively. Drones of Kerala possess a maximum length of the hind femur and hind tibia, followed by drones of Karnataka (2.93 and 4.57 mm). The metatarsus length and width were most significant in Karnataka, Tamil Nadu, Kerala, and Assam, with values ranging from 2.71 to 2.80 mm and 1.34 to 1.44 mm, respectively. The ratio of length and width of the metatarsus was the least in drones from Assam (1.89) and the highest in drones from Maharashtra (2.16). The size and width of the forewing vary from 13.20 to 13.73 and 4.16 to 4.37 mm, respectively.

Further, they noted that the length to width of the forewing was more or less uniform (3.11 to 3.18). The cubital index has been reported maximum in drones from Maharashtra (7.94), followed by Tamil Nadu (6.81). On the other hand, the lowest cubital has been observed in drones from Assam (4.49). They further reported that drones from Assam possess the maximum hamuli (25.67), whereas the least number has been reported from Maharashtra (21.00). Further, they noted that hairy patches were more on the cervix of male genitalia in drones from Karnataka than from Tamil Nadu.

Schlüns *et al.* (2003) reported a difference of about 7% and 13% in the wing length of small and large drones, respectively. Gençer and Firatli (2015) wrote that the forewing length of drones produced by a queen and by workers is about 10.8% and 2.4%, respectively.

Gençer and Firatli (2015) compared drones reared in queenright colonies (QRC) and egg-laying workers regarding different ages, reproductive capacities, and morphological characteristics. They concluded that depending upon comb type, there is a variation in drone length. Drones from queen-righted colonies were 17.0% heavier than drones from queen-less colonies, respectively. Drones from queen-righted colonies and egg-laying workers lost 18.8% and 13.3% of their initial weights during emergence, respectively. Drones from queen-righted colonies possess a significantly greater number of drones than those from egg-laying workers.

Furthermore, they determined the weight at different ages after emergence and compared drones from queens with drones from workers regarding reproduction and morphological characteristics. Berg (1991) studied drones from the queen and the weight of drones from workers and concluded that drones developed from the queen are heavier than drones from workers, with a weight of 260.8 mg and 151.8 mg, respectively. Similarly, Gençer and Firatli (2015) reported that drones from Queen possess an average weight of 274.3 mg. Furthermore, they concluded that heavier drones have larger mucus glands and seminal vesicles, which produce more spermatozoa than smaller drones. Mackensen (1955) reported that, on average, drones make 10.41×10^6, whereas Woyke (1962) said that drones produce 11.00×10^6 spermatozoa. Rinderer *et al.* (1985) speculated a difference in sperm number in Africanized and European drones, averaging 4.6×10^6 and 5.7×10^6 in single seminal vesicles of drones. Collins & Pettis (2001) reported that about 8.7×10^6 and 12.2×10^6 spermatozoa from drones emerged from cells uninfested with varroa and varroa-infested cells, respectively. Duay *et al.* (2002) reported that drones that emerge from uninfected cells and slightly varroa-infested cells produce 7.5×10^6 and 5.7×10^6 spermatozoa, respectively. Gençer and Firatli, 2015, reported that the spermatozoa number is 12.01×10^6 and 8.62×10^6 in drones from queens and workers.

Dois Vizinhos, 2018 showed that drone honey bees possess an average weight of 202.81 ± 17.84 mg at maturity. Drone honey bees possess a total length of about (15.39 ± 0.74), a length of abdomen (7.69 ± 0.68), a width of the abdomen (5.48 ± 0.29), a length of the wing (12.40 ± 0.66), and a width of the wing (3.83 ± 0.30). The average weight, area, and volume of the seminal vesicle were 1.80 ± 1.9 mg, 8.60 ± 2.92 mm^2, and 6.65 ± 3.31 mm^3, respectively. The average weight, area, and volume of the mucus gland were 12.60 ± 1.9 mg, 25.45 ± 8.59 mm^2, and 37.84 ± 18.12 mm^3, respectively.

Gazizova *et al.*, 2020 observed that probosci's length had a maximum value of 4.45 ± 0.05. Ruttner's method (2006) reported that drones possessed an average probosci's length from 3.81 ± 0.08 mm to 3.97 ± 0.03 mm, the cubital index ranges from 1.36 ± 0.07 to 1.45 ± 0.09, the right front wing's length ranges from 11.98 ± 0.49 to 12.08 ± 0.45 mm, the right front wing's width ranges from 3.72 ± 0.21 to 3.92 ± 0.23 mm, the third tergite's length ranges from 2.90 ± 0.12 to 3.51 ± 0.01 mm, the third tergite's width ranges from 6.40 ± 0.02 to 6.44 ± 0.04 mm, the third sternite's length ranges from 2.62 ± 0.03 to 2.66 ± 0.05 mm, the third sternite's width ranges from 4.52 ± 0.10 to 4.62 ± 0.07 mm, and tarsal index ranges from 49.59 ± 0.89 to $52.15 \pm 1.93\%$. Further, they noted variations in proboscis' length in drones from different regions being 3.90±0.12 (Kashaya), 3.88±0.10 (TalliYalan), 3.93±0.07 (Baysalyan), 3.90±0.10 (Kapova cave), 3.91±0.09 (Balatukay), 3.89±0.14 (Gadelgareevo), 3.92±0.08 (Kush-Alga bash).

DRONE'S REPRODUCTIVE SYSTEM METRIC REVIEW

Schlüns *et al.* (2003) found that wing length is positively phenotypically correlated with sperm numbers. Genetic correlation should also be considered in a breeding program instead of phenotypic correlation only. Abdominal length and width are relevant because reproductive organs are located there, which indicate the drone's growth and body development. Abdominal morphometrics shows reproductive potential and can be used as a selection criterion in regulating a drone's reproductive performance (Halak, 2012; Martins, 2014; Rodrigues, 2016). According to El-Kazafy and Abdulaziz (2013), *Apis mellifera jemenitica* and *Apis mellifera carnica* correlated emergence and sperm number at 14 days and concluded that there is a positive phenotypic correlation between body weight and: testicle size (0.99), seminal vesicle size (0.99), and mucus gland size (0.96).

Further, they concluded that the size of the seminal vesicle and mucus gland might be related to the size of the testicle with a positive phenotypic correlation of 0.99 and 0.97, respectively. Furthermore, they correlated the number of spermatozoa to the size of the testis, mucus gland, and seminal vesicle. The data obtained were 0.99, 0.99, and 0.97, respectively. Sperm volume can be influenced by environmental factors, reaching its peak in the spring and decreasing gradually as summer passes and autumn arrives. The genetic parameters of drone bees' morphometric traits and reproductive organs are fundamental in any apicultural production system.

The reproductive success of drones is dependent upon their size. According to Woyke (1962), sexually mature drones produce semen of about 1.50 and 1.75 mm^3 with about 11 million spermatozoa with a concentration of 7.5 million spermatozoa per 1 mm^3. According to Berg *et al.* (1997), smaller drones reared in worker cells possess certain disadvantages compared to large drones raised in drone cells. Rinderer *et al.* (1985) reported that Africanized drones contain fewer spermatozoa than European drones.

Ebadi & Gary (1980), Kaftanoglu & Peng (1982), and Harbo & Szabo (1984) reported that when a queen is inseminated instrumentally, they keep about 2.53 × 106, 3.83 × 106 and 3.2 × 106 spermatozoa in the spermathecae, whereas those who mate naturally possess 4.96 × 106, 4.54 × 106 and 5.5 × 106 spermatozoa, respectively. Further, in these studies, queens inseminated instrumentally possess about 4.57 × 106 and 3.61 × 106 sperm, whereas those mated naturally possess

6.13×10^6. According to Woyke (1979), a lower number of spermatozoa in instrumentally inseminated queens was caused by less favourable conditions than those of naturally mated queens. The queen kept in individual cages had only 2.736×10^6 spermatozoa in the spermatheca, while those kept in whole combs possessed about 4.690×10^6 spermatozoa. Schlüns *et al.* (2003) found 11.95×10^6 and 7.45×10^6 spermatozoa in large drones from queens and workers.

Also, queens kept in individual cages had only 1.8×10^6 spermatozoa in their spermatheca, while those moving freely in nucleus colonies had 5.2×10^6 spermatozoa, which is 2.9 times more (Woyke, 1989). Therefore in instrumentally inseminated queens in mating boxes like the naturally mated queens, the number of spermatozoa entering the spermathecae would be much higher.

CONCLUSION

Drone honey bee morphological features vary as per genomic possession (species-specific), protein formation and functionality, ecological conditions, food, temperature, etc. Drone morphometric characters are also related to mating capability, sperm production, sperm viability, etc. Additionally, drone morphometric characters reflect and also highlight queen honey bee genomic inheritance. The discussion cited in this chapter provides a comprehensive conclusion that drone morphometric characteristics vary per environmental cues.

REFERENCES

Bahrmann, R (1961) Ober den Bau des Begattungsschlauches von vier Apis-Arten. *Leipziger Bienenzig,* 75, 18-20.

(1991) Investigation on the rates of large and small drones at a drone congregation area. *Apidologie (Celle),* 22, 437-8.

Berg, S, Koeniger, N, Koeniger, G & Fuchs, S (1997) Body size and reproductive success of drones (Apis mellifera L). *Apidologie (Celle),* 28, 449-60.
[http://dx.doi.org/10.1051/apido:19970611]

Butler, CG (1957) The control of ovary development in worker honeybees (Apis mellifera). *Experientia,* 13, 256-7.
[http://dx.doi.org/10.1007/BF02157449] [PMID: 13447955]

(2001) Effect of varroa infestation on semen quality. *Am Bee J,* 141, 590-3.

de Groot, AP & Voogd, S (1954) On the ovary development in queenless worker bees (Apis mellifica L.). *Experientia,* 10, 384-5.
[http://dx.doi.org/10.1007/BF02160554] [PMID: 13210347]

Vizinhos, Dois Genetic aspects of morphometric traits and reproductive organs of africanized honey bee drones, Apis mellifera L. (Hymenoptera: Apidae)

Duay, P, De Jong, D & Engels, W (2002) Decreased flight performance and sperm production in drones of the honey bee (Apis mellifera) slightly infested by Varroa destructor mites during pupal development. *Genet Mol Res,* 1, 227-32.
[PMID: 14963829]

Abdou Taha, E-K & Alqarni, AS (2013) Morphometric and reproductive organs characters of apis mellifera jemenitica drones in comparison to apis mellifera carnica. *International Journal of Scientific & Engineering Research,* 4

Gazizova, N R, Galieva, Ch R & Tuktarov, V R (2019) A comprehensive morphological assessment of Apis mellifera drones in the Southern Ural. *Bulletin of the Bashkir State Agrarian University,* 21, 90-117.
[http://dx.doi.org/10.31563/1684-7628-2019-50-2-65-72]

Gençer, HV & Firatli, Ç (2005) Reproductive and morphological comparisons of drones reared in queenright and laying worker colonies. *J Apic Res,* 44, 163-7.
[http://dx.doi.org/10.1080/00218839.2005.11101172]

Jay, SC (1968) Factors influencing ovary development of worker honeybees under natural conditions. *Can J Zool,* 46, 345-7.
[http://dx.doi.org/10.1139/z68-052]

(1982) Effects of insemination on the initiation of oviposition in the queen honeybee. *J Apic Res,* •••, 21.

Koeniger, G & Koeniger, N (1990) Unterschiedliche Genese der Polyandrie bei Apis-Arten.*Verh Dtsch Zool Ges* Gustav Fischer Verlag, Stuttgart 618.

Koeniger, G, Koeniger, N, Mardan, M, Otis, G & Wongsiri, S (1991) Comparative anatomy of male genital organs in the genus Apis. *Apidologie (Celle),* 22, 539-52.
[http://dx.doi.org/10.1051/apido:19910507]

Gazizova, N R, Mannapov, A G & Sattarov, V N Morphological characterization of the Apis mellifera drones in the Southern Urals. *International AgroScience Conference.*

Pettis, JS, Higo, HA, Pankiw, T & Winston, ML (1997) Queen rearing suppression in the honey bee - evidence for a fecundity signal. *Insectes Soc,* 44, 311-22.
[http://dx.doi.org/10.1007/s000400050053]

Taha, RA (2008) Effect of varroasis on honey bee queen. M.Sc. Thesis. Fac. Agric. Kafr El-Sheikh, Univ., Egypt.

Ratnieks, FLW & Visscher, PK (1989) Worker policing in the honeybee. *Nature,* 342, 796-7.
[http://dx.doi.org/10.1038/342796a0]

Rinderer, TE, Collins, AM, Pesante, D, Daniel, R, Lancaster, V & Baxter, J (1985) A comparison of Africanized and European drones: weights, mucus gland and seminal vesicle weights, and counts of spermatozoa. *Apidologie (Celle),* 16, 407-12.
[http://dx.doi.org/10.1051/apido:19850405]

Ruttner, F (2006) *Biogeography and taxonomy of Honeybees.*Springer-Verlag, Berlin, Heidelberg.
[http://dx.doi.org/10.1007/978-3-642-72649-1]

Ruttner, F (1983) *Breeding technique and breeding selection.*Ehrenwirth, Munich.

Ruttner, F (1988) *Biogeography and Taxonomy of honey bees.*Springer, Berlin.

Ruttner, F & Hesse, B (1979) Breed-specific differences in ovarian development and oviposition in orphan workers of the honey bee Apis mellifera L. *Apidologie,* 12, 159-83.

Schlüns, H, Schlüns, EA, van Praagh, J & Moritz, RFA (2003) Sperm numbers in drone honeybees (*Apis mellifera*) depend on body size. *Apidologie (Celle),* 34, 577-84.
[http://dx.doi.org/10.1051/apido:2003051]

Viraktamath, S (2015) Morphometry and genitalia of drones of apis honey bee species from India. 10, 1057-67.

Simpson, H (1970) The male genitalia of Apis dorsata F. (Hymenoptera: Apidae). *Proc R Entomol Soc Londo,* A45, 169-71.

Visscher, PK (1989) A quantitative study of worker reproduction in honey bee colonies. *Behav Ecol*

Sociobiol, 25, 247-54.
[http://dx.doi.org/10.1007/BF00300050]

Visscher, PK (1998) Colony integration and reproductive conflict in honey bees. *Apidologie (Celle),* 29, 23-45.
[http://dx.doi.org/10.1051/apido:19980102]

Woyke, J (1962) Natural and artificial insemination of queen honeybees. *Bee World,* 43, 21-5.
[http://dx.doi.org/10.1080/0005772X.1962.11096922]

WOYKE, J (1979) Effect of the access of worker honeybees to the queen on the result of instrumental insemination. *J Apic Res,* 19, 136-43.

CHAPTER 3

The Development of the Drone Honey Bees: The Parthenogenesis

Abstract: The drone honey bee develops from unfertilized or fertilized eggs depending on the homozygosity of the sex alleles in inherited genomic content. In the honey bee colony, if the polyandrous queen honey bee mates with the drone honey bees of the neighbouring colonies, then the drones develop from the unfertilized eggs, confirming the haploid parthenogenesis. However, the mating of the queen with the drones of the same colony accelerates the feasibility of the development of drones, even from fertilized eggs. In the above-mentioned former case, the drones are known as the haploid drones, whereas in the latter case, the drones are referred to as the diploid drones. Generally, the diploid drones are removed by worker honey bees by recognizing the pheromones coated on the egg surface. The worker honey bees can remarkably distinguish the queen's drone eggs and the workers acting as pseudo-queens' drone eggs. The pseudo-queen develops if the colony is queen-less or the queen is not carrying the required reproductive potential and pheromonal emission. Drone development takes 24-25 days in total, with four distinct phases: egg, larval, pupal, and adult, with durations of 3 days, six days, 15-16 days, and about 1-3 months, respectively. The present chapter is attributed to the drone honey bees' developmental synchronicity, haploid inheritance, parthenogenesis, and patrilineal genomic contribution to the colony that influences colonial behaviour, productivity, life span, immunity, and others.

Keywords: Drone honey bee, Ecdysis, Holometabolous, Haploid parthenogenesis.

INTRODUCTION

The haploid/diploid drones (n/2n=16/32) of *Apis mellifera* (*A. mellifera*), a hymenopteran insect, develop following a typical holometabolous pathway of development with the inclusion of stages such as eggs, larvae, pupae, and adults. The inheritance is exclusively maternal as they develop from the unfertilized eggs in the natural populations (Figs. **1-6**). Under optimal environmental conditions, about 24–25 days are required for complete synchronous developmental events. Furthermore, the duration of development can vary depending on genomic possession, transcriptional expression, protein functionality, pheromonal communication, food resource availability, temperature, disease/pest infection/infestation, and other factors. About three days are required for the haploid eggs to hatch; after that, five larval instars develop. The larval phase undergoes moulting to convert to the pupa that completes organogenesis within an enclosed compartment walled with bee wax. After pupation, imagoes emerge by breaking the seal of the capped wax cells. For the first three days, the drones are fed by the worker honey bees; after that, they can take care of their diet. They prefer to feed on honey and stored pollen grains. The average life span of drones can be up to 55-90 days, which further varies as per environmental cues. The purpose of the male caste in the colony is to mate with queens of other honey bee colonies to contribute patrilineal genomic content. The drone's genomic contribution can influence the general behaviour, strength, productivity, disease resistance, and other characteristics of the honey bee colony.

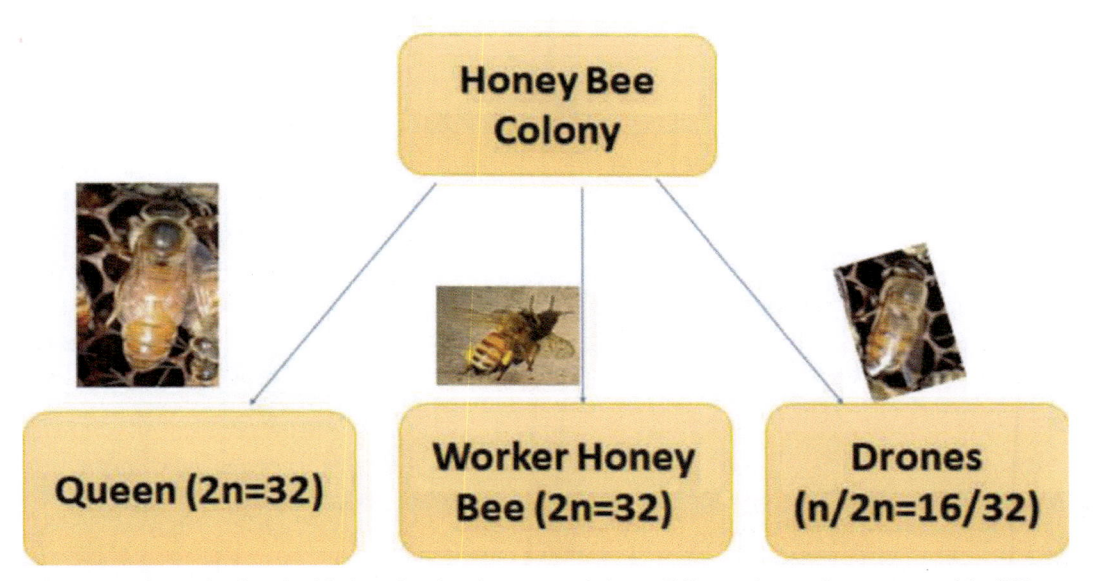

Fig. (1). Schematically elucidating the development of three different honey bee castes with different morphology, anatomy, physiology, proteomics, reproductive potential, life span, functionality, immunity, and others.

Furthermore, every honey bee colony needs to rear and maintain healthy drones because it is essential for successful queen bee mating. A polyandrous queen honey bee usually mates once with about 12–20 drones in her life span. A properly mated queen carries about 4.3 to 7 million spermatozoa in her spermathecae, sufficient to lay fertilized eggs during her reign in the colony (Rhodes, 2002). After mating, the queen stores sperm in her spermathecae, keeping it viable even after the drone that produced the male gametes dies (Klenk et al., 2004; Phiancharoen et al., 2004). A drone must find the queen in the drone congregation area and compete with hundreds of other drones from the same or different colonies to mate with her (Gary, 1963; Ruttner and Ruttner, 1972; Page, 1986; Koeniger, 1988; Berg et al., 1997). Hence, drone honey bees exhibit long anatomical and physiological adaptations for strong and forceful flight (Radloff et al., 2003).

A colony can raise a new queen if the existing queen is not mated correctly and does not carry sufficient sperm storage for egg laying. Generally, a queen with 3 million sperm after mating can be superseded within 12 months under commercial conditions, which is not acceptable for beekeeping as it accelerates swarming in the colony. A normal, reproductively active queen honey bee uses 2 million stored sperm to oviposit and fertilize eggs annually. A 16-day-old, reproductively healthy drone honey bee can produce about 5-10 million sperms. Further, the suitability of drones for mating decreases after 28 days of age.

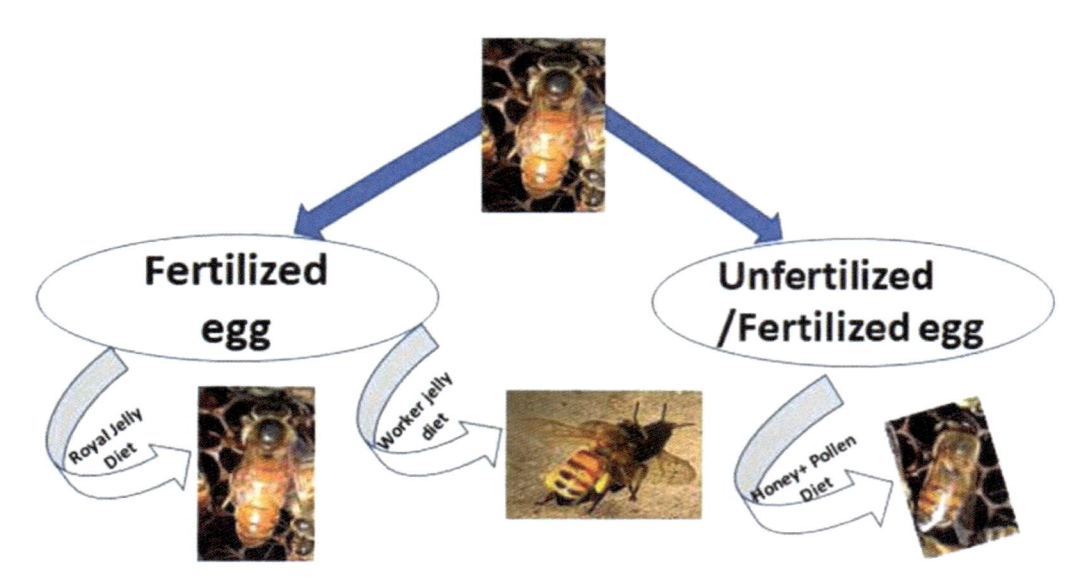

Fig. (2). Description of the polyandrous queen's ability to lay fertilized and unfertilized eggs, which can develop into female castes (the queen and the workers). The phenomenon of haploid parthenogenesis can be easily exemplified in the case of drone honey bees. In contrast, queen and worker honey bees witness polyphenism, the developmental plasticity of the same genomic contents expressed differentially per environmental cues.

In short, the purpose of the drone honey bee's life is the production of sperm and mating with the queen. Drones possess elaborative mating organs, powerful sense organs, comparatively larger eyes with numerous omatids, and long antennae with sensilla (Woyke and Ruttner, 1958; Koeniger, 1986; Koeniger et al., 1991; Koeniger and Koeniger, 1991, 2000; Seidl, 1980). Drones do not possess hypopharyngeal glands, wax glands, or other structures for food collection. Drones' heads are more prominent, their tongues are shorter, their mandibles are smaller, their mandibular glands are minor, and their stomachs are slenderer. Furthermore, no pollen basket is present on the meta-thoracic legs of drone honey bees (Snodgrass, 1956; Lensky et al., 1985).

DRONE DEVELOPMENT AND THE NEED FOR A PROTEIN-RICH DIET

The number of drones per colony varies as per the colony's strength. In other words, the more robust colony usually rears and maintains comparatively more drones than weaker colonies. The reason is the requirement of healthy worker bees feeding developing drones with honey, pollen, and gland secretions. Therefore, the low number of healthy nurse bees can challenge the rearing of drone bees in the colony. Furthermore, a continuous supply of fresh pollen is required to maintain the high population of drone honey bees as drones do not prefer to feed on the stored pollen grain.

Pollen stored near brood cells positively influences the drone rearing ability. A strong colony's daily requirement is 300–400 gm of high-quality pollen. If stored pollen is not available, in that case, pollen can be fed to the colony artificially. The conclusion drawn from an exploration indicates that pollen added to sugar syrup provides better nourishment and hence healthy growth to the drone brood than the drone larvae fed on the dry pollen grains. Additionally, it has been elucidated in the same study that colonies with a continuous supply of pollen exhibit comparatively better growth than colonies that do not receive a constant supply of pollen through foragers. Further, the pollen for the colony diet can be irradiated to ensure reduced infection and infestation from different diseases.

In the event of a fall in pollen supply, the drone honey bees can be evicted from the colony. With a shortage of pollen supply, a pollen-deficient diet is given to drones that eventually results in the weakening of the drones and, after that, they are forcefully evicted from the hive. The primary factor influencing the drone honey bee is the pollen grain supply. If the pollen grain supply is terminated in the colony, then the eviction of drones occurs. In contrast, if the pollen grain supply remains continuous to the colony, in that case, eviction of drones is reduced. The queen honey bee's absence also reduces eviction but not to that level of pollen grain supply (Rowlands and McLellan, 1987).

Fig. (3). A click of a hive section highlighting drone honey, which the blunt abdomen can recognize.

INFLUENCE OF ENVIRONMENTAL STRESSORS ON DRONE DEVELOPMENT

The drone honey bees are the reproductively active and obligatory sexual partners of the queen honey bee. Further, the presence of a healthy and high-quality drone influences the queen's fertility and the colony's productivity. The influence of stressors on drone fertility, survival, and physiology is limitedly known.

McAfee *et al.*, 2022 investigated the susceptibility of various abiotic stressors like cold stress, topical imidacloprid exposure, and exposure to a cocktail of pesticides on the drone honey bees. They had concluded that the drones were more sensitive to cold and imidacloprid exposure than the workers. After that, they carried out quantitative proteomics using the hemolymph of exposed drones and found that 34 proteins were differentially expressed in exposed drones compared to control. According to the researchers, drones represent a surprisingly higher baseline level of a putative stress response than workers do. Further, the concerned exploration demonstrated that the stress tolerance in the drone is fundamentally rewired, which indicates that susceptibility to stress is influenced by more factors than gene dose or allelic diversity.

Few reports are available in the scientific literature that attests that several factors influence drone fertility and development. McAfee *et al.* (2003) concluded that pesticide exposure and low temperatures adversely affect drone fertility. Fisher and Rangel, 2022, explored the influence of pesticide exposure on larval development and concluded that specific xenobiotics adversely impact drone development. Finally, Kairo *et al.* (2016) investigated the influence of fipronil on drone fertility.

Grass *et al.* (2018) concluded that drone honey bees are more sensitive to thiamethoxam exposure than workers. O'Donnell and Beshers (2004) concluded that the greater sensitivity of drones to environmental stressors could be due to the limited allelic diversity that comes with heterozygosity. Furthermore, a few reports show that the haploid susceptibility hypothesis does not support pathogenic infection in drones and workers (Ruiz-González and Brown, 2006), whereas some scientific studies support the hypothesis. The drone honey bees are susceptible to *Nosema*, leafcutter ants, wood ants, and buff-tailed bumble bees (Gerloff *et al.*, 2003; Vainio *et al.*, 2004; Baer *et al.*, 2005; Retschnig *et al.*, 2014), while Ruiz-González, *et al.*, 2006 elucidated that males are susceptible to *Crithidia*. According to the haploid susceptibility hypothesis, drones are more vulnerable to parasites and pathogens due to their haploid possession and reduced allelic diversity. Thiamethoxam and clothianidin pesticides have been reported to influence drone development and reduce drone fertility and life span (Friedli *et al.*, 2020; O'Donnell and Beshers, 2004).

Furthermore, McAfee *et al.* (2003) demonstrated that haploid drone honey bees were more susceptible to cold and pesticide stress than diploid worker bees. Even though they suggested that the haploid susceptibility hypothesis broadly applies to abiotic and biotic stressors, they further exposed developing drones and adults to agrochemical cocktails composed of pesticide-laced pollen patties. Still, they did not get consistent effects of pesticide treatment on the adults or the larvae. Researchers further detected that the drones express a higher level of putative stress response proteins than the workers. Additionally, they concluded that drones differentially regulate the same proteins in response to pesticide stress. For example, the hex110 expression protein is massively down-regulated in the drones compared to the workers.

Furthermore, according to them, the haploid susceptibility hypothesis does not explain why drones are sensitive to stress as they possess half the gene dose but express the higher basal level of stress response proteins. According to Hrassnigg and Crailsheim (2005), drones have a high amino acid reserve in hemolymph, which assists in expressing different proteins. One of the top five differentially expressed proteins was adenylate kinase, involved in cellular proliferation, energy homeostasis regulation, and AMP-induced cell signalling.

McAfee *et al.*, 2003 suggest that drone honey bees express high baseline expression proteins linked to oxidative damage, detoxification, temperature stress, DNA damage, and immune protein expression, which empower drones to combat various stressors. Further, they quantified about 654 protein groups related to stress response in drone honey bees. They hypothesized that worker honey bees exhibited better stress response, which can be due to post-translational modification of proteins to modulate the proteins' specific activities. They demonstrated that in drones, there is a differential expression of specific heat-shocked proteins than workers. In drones, there is a higher expression of HSP beta 1, HSP cognate 3, 97 kDa HSP, protein lethal(2) essential for life, and HSP70 Ab, and a putative glutathione-S-transferase and low expression for HSP60 and HSP10 compared to workers (Tan *et al.*, 2018). HSP60 and HSP10 facilitate the folding of proteins imported to mitochondria (Tan *et al.*, 2018; Kaufman *et al.*, 2003). There is a further decrease in production for HSP10 and HSP60 after imidacloprid treatment, which indicates that drones become more sensitive to heat stress after imidacloprid exposure.

Further, imidacloprid exposure can reduce proteins like HSP70 Ab and HSP cognate 3 (Koo *et al.*, 2015). Generally, drone honey bees express a high level of serpin 88Ea, trypsin-1-like, inositol-3-phosphate (I-3-P) synthase, and adenylate kinase (McAfee *et al.*, 2003). Further, they concluded that drones are more sensitive to cold stress and imidacloprid exposure; therefore, putative stress response proteins are highly expressed. The specific proteins are involved in detoxification, DNA repair, immunity, and oxidative stress. Some proteins are expressed more strongly in drones than in workers, like hex110 (a significant amino acid storage protein).

Fig. (4). Photograph of a hive section containing two honey bee castes: workers and drone honey bees. Worker honey bees can be observed performing different duties. Some adult drones and isolated capped drone pupae are also visible in the hive.

GENERAL DRONE DEVELOPMENT

Normally, haploid drones develop from unfertilized eggs laid by a queen or worker bees, whereas diploid drones develop from fertilized eggs if the queen mates with drones that share sex alleles. Usually, a diploid drone is eaten by workers within a few hours of laying. About 24 days are required for drone development. But, when drone cells are present on the periphery, the development of drones requires a longer time, *i.e.,* 25 days (Fukuda and Ohtani, 1977), which is most likely to happen due to variations in brood nest temperature. The drones solicit honey from workers and also take honey from honey cells. For the first ten days, drone development is completed within open cells, and during the specific developmental phase, larvae are cared for by workers. Capping occurs during the final 14 days of growth (Jay 1963). No diet is required during the prepupal and pupal stages (de Oliveira and Engels 2013). The workers feed the drones for the first three days of their adult life, but after that, they tend to feed themselves (Free, 1957). However, minimal exploration is available regarding the development of drones. The description regarding development is briefed below:

Drone Honey Bee Egg Stage

Queen and worker honey bees develop from fertilized eggs, whereas drone honey bees develop from unfertilized eggs. Further, diploid or haploid drones can develop from fertilized or unfertilized eggs laid by the queen or by the workers. Some reports claim that in honey bees, fertilized and unfertilized eggs do not differ in size (Henderson, 1991; Gencer and Woyke, (2006), whereas Reinhardt, 1960, claimed that both these eggs, including fertilized and unfertilized, vary in size over the season and between different colonies. Drones' eggs are extended and broader than worker eggs (Reinhardt, 1960). Although the size difference is relatively small, that could be why Henderson (1991) did not find a significant difference between male and female eggs.

The length of a honey bee egg is 1.49±0.12 (range 1.12–1.85) mm, the width is 0.35±0.02 (range 0.30–0.40) mm, the volume is 0.10±0.02 (range 0.06–0.15) cubic mm (Woyke J., 1998). The dimension of the egg changes markedly with growth. It does not increase immediately after laying eggs but before hatching. The egg dimension varies between different subspecies and between queens of the same subspecies (Woyke, 1998; Taber and Roberts, 1963; Gencer and Woyke, 2006). Further, the weight of an egg varies from 0.12 to 0.22 mg (Taber and Roberts, 1963).

Harbo and Bolten (1981) reported that drone eggs develop slightly slower than those worker eggs at the same temperature (as reported by Harbo and Bolten, 1981). Male larvae develop approximately three hours later than worker larvae at

34.8 °C, with a mean hatching time of 71.4±1.2 h (Mean ± SD). According to Mackasmiel and Fell (2000), the respiration rate of two kinds of eggs is temperature-dependent, but they could not detect any difference in consumed oxygen.

Larval Stage of the Drone Honey Bee

Unfertilized or fertilized eggs hatch into larvae, feeding on royal jelly for the first three days and then on drone jelly. A total of five larval instars are present in the development of drones.

Drone larvae attain 1.8–2.6 times the weight of worker larvae. The weight of a freshly emerged European honey bee larva is about 277–290 mg (Schneider and Drescher, 1987; Duay *et al.*, 2003), whereas the weight of drone larvae at the time of cell sealing is 262–419 mg (Strauss, 1911; Nelson *et al.*, 1924; Stabe, 1930; Bishop, 1961; Jay, 1963; Wang, 1965; Thrasyvoulou and Benton, 1982).

A few available reports claim the biochemical characteristics of the drone larvae. The lipid content in the worker larva increases with its development and further decreases during metamorphosis (Straus, 1911; Melampy *et al.*, 1940; Cantrill *et al.*, 1981). According to Straus (1911), the pattern of glycogen accumulation in drones is similar to that of worker larvae. In drones, absolute and relative glycogen content decrease towards the imaginal stage. Straus (1911) noted that drone larvae accumulate more lipids than worker larvae. Melampy *et al.* (1940) further determined the lipid content of drone larvae at the time of cell sealing to be 5.4% of fresh weight (21.3% of dry weight).

Drone larvae are fed on secretions from nurse workers' hypopharyngeal glands, and mandibular and postcephalic glands (Brouwers, 1982; Knecht and Kaatz, 1990; Lensky and Rakover, 1983). A minor protein diet is obtained by feeding on natural pollen grains. Babendreier *et al.* (2004) studied the amount of pollen fed to dwarf larvae, which were lighter in weight than normal drone larvae. They further found that drone larvae were 39% heavier than worker larvae and consumed 36% more pollen than worker larvae. Simpson (1955) found comparatively low pollen consumption in worker larvae. Drone larvae are fed on pollen with a minimal quantity of diet for the first three days (Matsuka *et al.*, 1973). Food provided to young drones of 1-3 days of age is milky white in appearance, whereas the older larvae are fed a yellow-brown diet due to the addition of honey and pollen. In later larval stages, food contains less protein and fat but more carbohydrates (Haydak, 1957a, 1970). During early development, the Drone larval diet is similar to worker larvae for protein and carbohydrate content. Drone larval jelly is changed after 108 h, compared to worker jelly, which is changed after 84 hrs (Brouwers, 1984; Brouwers *et al.*, 1987). This could be the reason for the delayed development of drones compared to workers (Stabe, 1930). Rhein (1951) considered the difference between worker and drone jelly of little relevance, as he was able to raise one drone larvae to adult and another to the pupal stage with worker jelly.

There is an increase in protein content with the growth of larvae, but the relative amount of protein per body weight decreases (Straus, 1911; Melampy et al., 1940; Imdorf et al., 1998) or directly (Hepburn et al., 1979;Kunert and Crailsheim, 1987). In drone larvae, the maximum amount of protein is reached when the cell is sealed, and after that, the amount of protein remains relatively constant (Straus, 1911; Melampy et al., 1940).

The high level of glycogen and lipids in drone larvae' diet can increase their dry weight. These biochemicals provide molecules for building the imaginal organs and energy for the process (Melampy et al., 1940). Drone larvae need several carbohydrates of about 98.2 mg for a drone (Rortais et al., 2005). Allen (1959) reported that drone larvae consume more oxygen than worker larvae, which can be due to the growth rate of drone larvae. However, Allen's study had not been supported by Stabe (1930), who had demonstrated that drone larval growth is slower than the worker larva.

Further, drone jelly is a rich source of about 20 amino acids, comprising 8.7% Free State and 15.9% exogenous amino acids. The essential amino acids of drone jelly include threonine, valine, methionine, isoleucine, leucine, phenylalanine, lysine, histidine and tryptophan. The major proportionality of different amino acids contains glutamic acid (6.5% of all amino acids), leucine and aspartic acid (3.6% each), proline (3.4%), lysine (2.9%), valine (2.3%) and alanine (2.1%) which make up about 60% of all amino acids. Further, in drone jelly lipids including triglycerides, free fatty acids, fatty acid esters and decenoic acids. The proportionality of different lipids includes saturated acids (40%), of which palmitic and stearic acids are the most abundant, and 50% of the contents were monounsaturated acids. In contrast, oleic acid constitutes 32.3% of this group. In drone jelly, the percentage of glucose and fructose was found 68.3% and 11.4%, respectively. Drone jelly includes vitamin Thiamine, Riboflavin, Nicotinic acid, Choline, Pantothenic acid, Pyridoxine, Retinol, Beta carotene, alpha-tocopherol, and Calciferol (Sidor and zugan, 2020)

The sting and certain external rudiments of genital organs can be used to determine the sex of larvae (Michaelis, 1900; Zander, 1900; Zander, Loschel, & Meier, 1916). During the larval phase of development, sting rudiments and the lobes of the rudiments of external drone genital organs remain covered within the larval cuticle, making it challenging to identify. In three-day-old larvae, one white imaginal disc is visible on the ventral side of the twelfth segment, whereas, in females, no such disc is visible. The sex of larvae can not be determined until the third day. In the fourth instar, sex can be easily discriminated by the characteristic greenish-bluish fluorescence of all imaginal discs on the ventral side of the female larva's tenth, eleventh, and twelfth segments; in males, it is present only on the ventral side of the twelfth segment. This fluorescence further decreased gradually in subsequent days (Woyke, 1963).

The Pupal Development in the Drone Honey Bee

The larva undergoes moulting to convert to a pupa. The pupal phase is completed within the sealed cell, and the pupal phase is primarily responsible for organogenesis. Furthermore, the concentration of protein in hemolymph varies with the pupal stage. The nitrogen content of drone pupae is higher by a factor of 2.6 than that of workers (Straus, 1911). About 65–97.5 mg of protein, derived from 325–487.5 mg of pollen, is needed to develop drone pupae.

The Adult Drone Honey Bee

The adult honey bee develops from a pupa after proper development. The adult drone honey bee body size varies depending upon which, wax cell is used to raise drones. The size of haploid drones developed in worker cells is smaller than those grown in drone cells. Such drones are called "dwarf drones" due to their small size. Such drones produce comparatively fewer spermatozoa than larger drones. But when calculated based on the weight per gram mass of the drone body dwarf drones produce 20% more spermatozoa than large drones (Schlüns *et al.*, 2003). Therefore, workers need to invest less in raising small drones, which produce comparatively more spermatozoa (Schlüns *et al.*, 2003). As workers grow drones in worker cells under specific circumstances, the more costly rearing of large drones in drone cells must be advantageous in some other aspects. For example, it could be that the larger sperm volume of a large drone increases its representation in the form of the queen's offspring. Another consideration is that a large-sized drone may have an advantage because it can compete for the queen during mating. As drone size is controlled by the wax cells used to raise drones, colony-level selection should favour the drone size that is optimal for fitness at the colony level (Kraus *et al.* (2004).

The life Span of Adult

Drones varies, as drones produced in the summer are short-lived, whereas autumn-reared drones are long-lived. That can be due to decreased flight activity (Fukuda and Ohtani, 1977). Generally, drone honey bees are evicted from their colonies in the autumn, so they do not live to their total expectancy (Mindt,

1962;Morse et al., 1967; Free and Williams, 1975; Colonello and Hartfelder, 2003).

In *Apis mellifera* drones, the mucus is pressed into the endophallus, facilitating comparatively stable connections between drones and the queen during mating (Koeniger and Koeniger, 1991). According to Colonello and Hartfelder (2003), drones' mucus gland protein helps stimulate oogenesis and oviposition, while other biochemicals induce mechanical stimulation effects that help in everting endophallus in the queen's bursa copulatrix (Koeniger, 1976, 1981). *Apis mellifera* drones produce more mucus protein and spermatozoa (10–12 million) than *Apis florea*, which makes little mucus and a smaller number of spermatozoa (0.44 million).

Tozetto et al. (1995) after eclosion juvenile hormone synthesis increases to a maximum in 9-day-old drones and then decreases. Furthermore, they reported that the application of JH III promoted the flight activity of drones.

Panzenbock and Crailsheim (1997) concluded that freshly emerged drones possess more glycogen than same-aged workers, with values of 0.55– 0.59 mg, i.e. 0.22% of fresh weight, and workers, 0.13 mg, i.e. 0.12% of fresh weight. Berger et al. (1997) described an increase in protein content between 5 d and 12 d old drones. The protein content accumulates more in the thorax and abdomen due to flight muscle location and sexual organ location (Crailsheim et al., 1997).

The age-related protein content of haemolymph has been reported in the case of drone honey bees (Sinitzky and Lewtschenko, 1971; Trenczek et al., 1989; Cremonez et al., 1998). The concentration of free amino acids in haemolymph depends significantly on the age of the drones (Crailsheim and Leonhard, 1997; Leonhard and Crailsheim, 1999). In drones, the free amino acid concentration is the highest at the age of 5 days old and decreases after that. In drone honey bees, the maximum concentration is 90.5 ± 10.98 nmol/μL (Hrassnigg et al., 2003a). In the drone caste, the predominant amino acid is proline present at concentrations of 31–52% (Crailsheim and Leonhard, 1997; Leonhard and Crailsheim, 1999). Proline is used for oxidative metabolism at a higher rate than phenylalanine or leucine (Berger et al., 1997).

Beutler (1937) determined the average sugar concentration in the hemolymph in drones at 1–2%. Sugars in the hemolymph of drones were glucose, fructose, trehalose, and sucrose (Alumot et al., 1969). In addition, drones consume pollen during the first few days of emergence (Szolderits and Crailsheim, 1993).

In drones, as in workers, the level of proteolytic activity in the midgut corresponds to the amount of pollen present (Szolderits and Crailsheim, 1993).

But in drones, the level of this activity was found to be lower than in workers (Free, 1957; Mindt, 1962; Crailsheim, 1991, 1992; Hrassnigg et al., 2003b). The information on pollen consumption, digestive enzyme activity, and transfer of proteinaceous jelly in workers and drones underline the principal function of workers in digesting food for the colony, and they show that drones rely on this pre-processed food.

Drones older than two days contribute to active heat production (Harrison, 1987; Kovac and Stabentheiner, 2002). Drones are ectothermic between 5 and 20 degrees Celsius but endothermic between 25 and 35 degrees Celsius (Cahill and Lustick, 1976; Norbert Hrassnigg, Karl Crailsheim, 2005). The drones consume their maximum food before the flight.

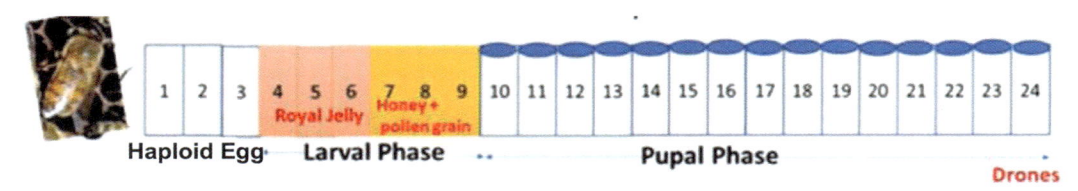

Fig. (5). The description of the drone honey bee's development from haploid eggs includes three days of egg life, six days of the larval phase, and 15 days of the pupal developmental stage.

Drone honey bees do not participate in foraging, colony maintenance, or disease. They believe the only purpose in their lives is to mate with virgin queens. The queen mates with 6–27 drones during the nuptial flight. After mating, drones die shortly. Drones can improve quality by contributing paternal genomic content to their subsequent progeny.

Predation influences the longevity of drones, and the queen's quality affects the construction and conversion of drone cells to worker cells through pheromones of the queen's mandibular glands. Additionally, drone numbers depend upon factors like colony size, time of year, and the number of drone cells. According to Fukuda and Othani, in 1977, about 50-56% of cells had the potential to develop into drones. Egg laying is regulated not only by queen honey bees but also by worker bees who destroy and eat drone eggs, larvae and occasionally pupae (Currie, 2015). Further, the drone brood laid at the periphery, making it more susceptible to chilling than worker bees. Also, honey bee colonies used for drone brood rearing should have a continuous supply of nectar or sugar syrup, pollen or pollen supplements. Large colonies can have about 1500 adult drones, and further strong colonies can even accept foreign drones, which can be introduced artificially or come from neighbouring colonies. Further, queenless colonies take more drones than queen-right colonies but with an equal survival rate (Currie, 1987).

Caste	Number in Colony	Egg Type	Develop-ment (Days)	Life Span	Duties	Image
Queen	1	Fertilized egg	16	1-3 years	Repro-duction Queen pheromone secretion	
Drone	100-1000	Unfertilized/fertilized egg	24	2month-6 months (Dependence on mercy of worker bees)	Repro-duction	
Worker	20000-50000	Fertilized egg	21	1 month-6 month	Foraging, temperature regulation, honey production, pollen storage, brood rearing, wax comb construction, ventilation, temperature regulation, protection, queen	

Fig. (6). Comparatively, the number, development, life span, and duties of different castes of honey bees in the colony

THE TEMPERATURE PREFERENCE OF ADULT DRONES

Further, adult drones prefer temperatures of about 35°C in the hive, and thermal preference varies according to age (Ohtani and Fukuda, 1977). The warmer part of the hive is preferred by younger drones, whereas older drones prefer the comparatively cooler part of the hive.

Krystyna and Tofilski, 2020 reported that high temperatures during the early stage promote the production of heavier drones. The body mass of drones can be enhanced by keeping them at changing temperatures rather than constant temperatures. Further, they concluded that there had been an increase in body mass when drones shifted from 35 °C to 32 °C during the post-capping period. The drone reared at a constant temperature of 35 °C possessed a lower body mass. They speculated that drones reared at changing temperatures exhibited less mortality and superior quality. They confirmed that the body mass of drones became highest at five days. Subsequently, body mass starts decreasing until the end of life.

There are various factors that influence the temperature of nests, like starvation, an inadequate number of workers, or diseases (Jaycox 1961; Kronenberg and Heller 1982; Fahrenholz *et al.* 1989; Starks *et al.* 2000; Kleinhenz 2003; Jones *et al.* 2005; Fehler *et al.* 2007; Stabentheiner *et al.* 2010; Campbell *et al.* 2010;

Rousseau and Giovenazzo 2016). The drones are reared at the periphery of the nest and are exposed to more temperature fluctuations. Drones raised at 32 °C possess more enormous reproductive organs and more viable spermatozoa, but a comparatively lower volume of semen than drones reared at 35 °C (Czekoska et al. 2013). Lower-temperature drones are larger and heavier (Szentgyörgyi et al., 2018; Czekoska et al., 2019). The large drones produce more spermatozoa, start mating flight earlier, show a higher probability of mating with the queen, and live longer (Schlüns et al. 2003; Gençer and Firatli 2005; Koeniger et al. 2005; Rueppell et al. 2006; Couvillon et al. 2010; Slone et al. 2012; Taha and Alqarni 2013; Czekoska et al. 2019). Therefore, comparatively larger drones are more advantageous to the colony than smaller ones with the overall mating success. Honey bee colonies influence the reproductive quality of the drone honey bees based on the larval rearing feature of nurse bees. In other words, better-reared drone bees are of superior quality than the ignored ones (Boucher and Schneider 2009; Slone et al. 2012).

Furthermore, drones developed at a constant higher temperature (35°C) are smaller and evicted more frequently than the drones developed at a lower temperature (32°C) (Szentgyörgyi et al., 2018; Czekoska et al., 2019). Moreover, the quality of drones influences queens' longevity and fertility (Brutscher et al., 2019; Metz and Tarpy, 2019; Rangel and Fisher, 2019). Finally, the body mass of adult drones decreases with age, independent of environmental factors (Gençer and Firatli 2005; Metz and Tarpy 2019).

A few more reports confirmed that the development of honey bees is influenced by the temperature of the nest (Seeley 2014; Abou-Shaara et al. 2017). In the outer area of the hive, temperatures of about 33°C and 36°C are maintained (Kleinhenz 2003; Jones and Oldroyd 2006; Seeley 2014), whereas, in the centre of a nest, a constant temperature close to 35 °C is maintained. Mainly worker-brood develops in the centre, whereas at the periphery with lower temperatures, mainly the drone brood develops (Seeley and Morse 1976; Levin and Collison 1990; Kleinhenz 2003; Seeley 2014; Li et al. 2016; AbouShaara et al. 2017).

Drones reared at 32°C produce lower numbers of sperm but more viability. The environment also influences drone body mass after emergence (Szentgyörgyi et al., 2017; Czekońska et al., 2019). The body mass of newly emerged males depends upon feeding, the number of attending workers, and nest temperature (Haydak 1970; Free and Williams 1975; Hrassnigg and Crailsheim 2005; Czekońska et al. 2015; Szentgyörgyi et al. 2016). At a later age, body mass is influenced by colony size, presence of a queen, food availability, and health status (Wharton et al. 2007, 2008; Boes 2010; Mazeed 2011; Czekońska et al. 2019). The body mass of the drones decreases after the fifth day of their lives. Lower body mass in drones allows the drones to spend more time in the air before returning to the colony (Harrison 1986). It is further assessed that drones reared at different temperatures change their body mass in variable ways. Drones nourished at 35°C are lighter and comparatively increase their body mass less post-emergence (Krystyna and Tofilski, 2020).

The worker honey bees change the rearing temperature from higher at an earlier stage to lower at a later stage of development to obtain heavier and better surviving drones. Generally, the drone brood is located at the edge of a hive. The specific location provides optimal conditions for the development of drones (Seeley and Morse 1976; Jaycox 1961; Jones et al. 2005). A few explorations indicated that at the nest periphery, the brood temperature is more variable, as it is influenced by conditions outside the nest (Kraus et al. 1998; Jones et al. 2004). Further, temperature fluctuations are exceptionally high between day and night (Kronenberg and Heller 1982). Although the diurnal temperature change is smaller during the day, it can reach up to 1.5 °C (Kronenberg and Heller 1982). Some reports indicate that the temperature of drones is lower and more variable than worker honey bees (Levin and Collison 1990; Li et al. 2016; Abou-Shaara et al. 2017). Low rearing temperature promotes the formation of a heavier body mass in drones and is more conducive to the formation of spermatozoa in the post-capping period. According to certain reports, during the first two days of drone emergence, sperms are produced and stored within the testes (Bishop 1920; de Oliveira and Engels 2013). The drone sperm survives better at lower temperatures (Woyke, 1962, 1963; Woyke, 1965; Stürup et al., 2013).

FACTORS INFLUENCING DRONE CELLS AND LIFE SPAN

Production of drones is influenced by multiple factors like colony strength, queen presence, time of year, and food availability (Free and Williams1975; Whartonetal. 2007,2008; Boes2010; Smith et al., 2015). Further, colonies influence the reproductive quality of drones (Wharton et al. 2008; Boes 2010). Under unfavourable environmental conditions, drones can be selectively evicted from colonies (Seeley and Mikheyev 2003; Hrassnigg and Crailsheim 2005; Czekońska et al. 2015).

The lifespan of drones varies, as indicated by various explorations. The drone life span varies according to season, with the average drone life span being 13 days in the summer and 38 days in the autumn. Further, the survival rate of drones is inversely related to the number of flights (Fukuda and Othani, 1977). The factors that create hindrances in insect flight increase the longevity of drones, including cool temperatures, high winds, or overcast skies. A colony has to invest more to nurse a drone honey bee than a worker bee (Haydak, 1970).

CONCLUSION

Concluding remarks from the above discussion include that drone development is influenced by temperature, pollen supply, the force of nurse bees, pheromonal, genomic content, and other factors. The development of diploid/haploid and large/dwarf drones differs according to ploidy level and wax cell size. The colonies maintain healthy drones to contribute to patrilineal genomic content, but under stress conditions, the colonies even evict the drones.

REFERENCES

Abou-Shaara, HF, Owayss, AA, Ibrahim, YY & Basuny, NK (2017) A review of impacts of temperature and relative humidity on various activities of honey bees. *Insectes Soc,* 64, 455-63.
[http://dx.doi.org/10.1007/s00040-017-0573-8]

McAfee, A, Metz, BN & Milone, JP (2022) Drone honey bees are disproportionately sensitive to abiotic stressors despite expressing high levels of stress response proteins. *Commun Biol,* 5, 141.
[http://dx.doi.org/10.1038/s42003-022-03092-7]

Len, MD (1959) Respiration rates of larvae of drone and worker honey bees, Apis mellifera L. *J Econ Entomol,* 52, 399-402.

Alumot, E, Lensky, Y & Holstein, P (1969) Sugars and trehalase in the reproductive organs and hemolymph of the queen and drone honey bees (Apis mellifica L. var. Ligustica spin.). *Comp Biochem Physiol,* 28, 1419-25.
[http://dx.doi.org/10.1016/0010-406X(69)90579-9]

Babendreier, D, Kalberer, N, Romeis, J, Fluri, P & Bigler, F (2004) Pollen consumption in honey bee larvae: a step forward in the risk assessment of transgenic plants. *Apidologie (Celle),* 35, 293-300.
[http://dx.doi.org/10.1051/apido:2004016]

Baer, B, Krug, A, Boomsma, JJ & Hughes, WOH (2005) Examination of the immune responses of males and workers of the leaf-cutting ant Acromyrmex echinatior and the effect of infection. *Insectes Soc,* 52, 298-303.
[http://dx.doi.org/10.1007/s00040-005-0809-x]

Berg, S, Koeniger, N, Koeniger, G & Fuchs, S (1997) Body size and reproductive success of drones (Apis mellifera L). *Apidologie (Celle),* 28, 449-60.
[http://dx.doi.org/10.1051/apido:19970611]

Berger, B, Crailsheim, K & Leonhard, B (1997) Proline, leucine and phenylalanine metabolism in adult honeybee drones (Apis mellifica carnica Pollm). *Insect Biochem Mol Biol,* 27, 587-93.
[http://dx.doi.org/10.1016/S0965-1748(97)00034-9]

Beutler, R (1936) Über den Blutzucker der Bienen. *J Comp Physiol A Neuroethol Sens Neural Behav Physiol,* 24, 71-115.
[http://dx.doi.org/10.1007/BF00340968]

Bishop, GH (1961) Growth rates of honey bee larva. *J Exp Zool,* 146, 11-20.
[http://dx.doi.org/10.1002/jez.1401460104]

Bishop, GH (1920) Fertilization in the honey-bee. I. The male sexual organs: Their histological structure and physiological functioning. *J Exp Zool,* 31, 224-65.
[http://dx.doi.org/10.1002/jez.1400310203]

Boes, KE (2010) Honeybee colony drone production and maintenance in accordance with environmental factors: an interplay of queen and worker decisions. *Insectes Soc,* 57, 1-9.
[http://dx.doi.org/10.1007/s00040-009-0046-9]

Boes, KE (2010) Honeybee colony drone production and maintenance in accordance with environmental factors: an interplay of queen and worker decisions. *Insectes Soc,* 57, 1-9.
[http://dx.doi.org/10.1007/s00040-009-0046-9]

Boucher, M & Schneider, SS (2009) Communication signals used in worker–drone interactions in the honeybee, Apis mellifera. *Anim Behav,* 78, 247-54.
[http://dx.doi.org/10.1016/j.anbehav.2009.04.019]

Brouwers, EVM (1984) Glucose/fructose ratio in the food of honeybee larvae during caste differentiation. *J Apic Res,* 23, 94-101.
[http://dx.doi.org/10.1080/00218839.1984.11100616]

Brouwers, EVM, Ebert, R & Beetsma, J (1987) Behavioural and physiological aspects of nurse bees in relation to the composition of larval food during caste differentiation in the honeybee. *J Apic Res,* 26, 11-23.
[http://dx.doi.org/10.1080/00218839.1987.11100729]

Brutscher, LM, Baer, B & Niño, EL (2019) Putative Drone Copulation Factors Regulating Honey Bee (*Apis mellifera*) Queen Reproduction and Health: A Review. *Insects,* 10, 8.
[http://dx.doi.org/10.3390/insects10010008] [PMID: 30626022]

Cahill, K & Lustick, S (1976) Oxygen consumption and thermoregulation in Apis mellifera workers and drones. *Comp Biochem Physiol A Comp Physiol,* 55, 355-7.
[http://dx.doi.org/10.1016/0300-9629(76)90060-8] [PMID: 9250]

Campbell, J, Kessler, B, Mayack, C & Naug, D (2010) Behavioural fever in infected honeybees: parasitic manipulation or coincidental benefit? *Parasitology,* 137, 1487-91.
[http://dx.doi.org/10.1017/S0031182010000235] [PMID: 20500914]

Colonello, NA & Hartfelder, K (2003) Protein content and pattern during mucus gland maturation and its ecdysteroid control in honey bee drones. *Apidologie (Celle),* 34, 257-67.
[http://dx.doi.org/10.1051/apido:2003019]

Colonello, NA & Hartfelder, K (2003) Protein content and pattern during mucus gland maturation and its ecdysteroid control in honey bee drones. *Apidologie (Celle),* 34, 257-67.
[http://dx.doi.org/10.1051/apido:2003019]

Couvillon, MJ, Hughes, WOH, Perez-Sato, JA, Martin, SJ, Roy, GGF & Ratnieks, FLW (2010) Sexual selection in honey bees: colony variation and the importance of size in male mating success. *Behav Ecol,* 21, 520-5.
[http://dx.doi.org/10.1093/beheco/arq016]

Crailsheim, K, Jin, P, Pfeiffer, KJ & Pabst, MA (1997) Der Proteingehalt junger Drohnen der Honigbiene (Apis mellifera L.). *Mitt Dtsch Ges Allg Angew Entomol,* 11, 167-8.

Crailsheim, K, Jin, P, Pfeiffer, KJ & Pabst, MA (1997) Der Proteingehalt junger Drohnen der Honigbiene (Apis mellifera L.). *Mitt Dtsch Ges Allg Angew Entomol,* 11, 167-8.

Crailsheim, K & Leonhard, B (1997) Amino acids in honeybee worker haemolymph. *Amino Acids,* 13, 141-53.
[http://dx.doi.org/10.1007/BF01373212]

Crailsheim, K, Schneider, LHW, Hrassnigg, N, Bühlmann, G, Brosch, U, Gmeinbauer, R & Schöffmann, B (1992) Pollen consumption and utilization in worker honeybees (Apis mellifera carnica): Dependence on individual age and function. *J Insect Physiol,* 38, 409-19.
[http://dx.doi.org/10.1016/0022-1910(92)90117-V]

Cremonz, TÂNM, De Jong, D & Bitondi, MÁRMG (1998) Quantification of hemolymph proteins as a fast method for testing protein diets for honey bees (Hymenoptera: Apidae). *J Econ Entomol,* 91, 1284-9.
[http://dx.doi.org/10.1093/jee/91.6.1284]

Czekońska, K, Chuda-Mickiewicz, B & Chorbiński, P (2013) The effect of brood incubation temperature on the reproductive value of honey bee (*Apis mellifera*) drones. *J Apic Res,* 52, 96-105.

[http://dx.doi.org/10.3896/IBRA.1.52.2.19]

Czekońska, K, Chuda-Mickiewicz, B & Samborski, J (2015) Quality of honeybee drones reared in colonies with limited and unlimited access to pollen. *Apidologie (Celle)*, 46, 1-9.
[http://dx.doi.org/10.1007/s13592-014-0296-z]

Czekońska, K, Szentgyörgyi, H & Tofilski, A (2019) Body mass but not wing size or symmetry correlates with life span of honey bee drones. *Bull Entomol Res*, 109, 383-9.
[http://dx.doi.org/10.1017/S0007485318000664] [PMID: 30205847]

de Oliveira, TS & Engels, W (2013) Classification of substages in preimaginal development of honey bee drones (Hymenoptera: Apidae). *Entomol Gen*, 287-93.
[http://dx.doi.org/10.1127/entom.gen/34/2013/287]

Duay, P, Jong, DD & Engels, W (2003) Weight loss in drone pupae (*Apis mellifera*) multiply infested by *Varroa destructor* mites. *Apidologie (Celle)*, 34, 61-5.
[http://dx.doi.org/10.1051/apido:2002052]

Sidor, E & Dżugan, M (2020) Drone Brood Homogenate as Natural Remedy for Treating Health Care Problem: A Scientific and Practical Approach. *Molecules*, 25, 5699.
[http://dx.doi.org/10.3390/molecules25235699] [PMID: 33287191]

Fahrenholz, L, Lamprecht, I & Schricker, B (1989) Thermal investigations of a honey bee colony: thermoregulation of the hive during summer and winter and heat production of members of different bee castes. *J Comp Physiol B*, 159, 551-60.
[http://dx.doi.org/10.1007/BF00694379]

Fehler, M, Kleinhenz, M, Klügl, F, Puppe, F & Tautz, J (2007) Caps and gaps: a computer model for studies on brood incubation strategies in honeybees (Apis mellifera carnica). *Naturwissenschaften*, 94, 675-80.
[http://dx.doi.org/10.1007/s00114-007-0240-4] [PMID: 17443307]

Fisher, A, II & Rangel, J (2018) Exposure to pesticides during development negatively affects honey bee (Apis mellifera) drone sperm viability. *PLoS One*, 13, e0208630.
[http://dx.doi.org/10.1371/journal.pone.0208630] [PMID: 30543709]

Free, JB (1957) The food of adult drone honeybees (Apis mellifera). *Br J Anim Behav*, 5, 7-11.
[http://dx.doi.org/10.1016/S0950-5601(57)80038-0]

Free, JB & Williams, IH (1975) Factors determining the rearing and rejection of drones by the honeybee colony. *Anim Behav*, 23, 650-75.
[http://dx.doi.org/10.1016/0003-3472(75)90143-8]

Friedli, A, Williams, GR, Bruckner, S, Neumann, P & Straub, L (2020) The weakest link: Haploid honey bees are more susceptible to neonicotinoid insecticides. *Chemosphere*, 242, 125145.
[http://dx.doi.org/10.1016/j.chemosphere.2019.125145] [PMID: 31678852]

Fukuda, H & Ohtani, T (1977) Survival and life span of drone honeybees. *Popul Ecol*, 19, 51-68.
[http://dx.doi.org/10.1007/BF02510939]

Fukuda, H & Ohtani, T (1977) Survival and life span of drone honeybees. *Popul Ecol*, 19, 51-68.
[http://dx.doi.org/10.1007/BF02510939]

Gary, NE (1963) Observations of mating behaviour in the honeybee. *J Apic Res*, 2, 3-13.
[http://dx.doi.org/10.1080/00218839.1963.11100050]

Gençer, HV & Woyke, J (2006) Eggs from *Apis mellifera caucasico* laying workers are larger than from queens. *J Apic Res*, 45, 173-9.
[http://dx.doi.org/10.1080/00218839.2006.11101344]

Gençer, HV & Fıratlı, Ç (2005) Reproductive and morphological comparisons of drones reared in queenright and laying worker colonies. *J Apic Res*, 44, 163-7.
[http://dx.doi.org/10.1080/00218839.2005.11101172]

Gerloff, CU, Ottmer, BK & Schmid-Hempel, P (2003) Effects of inbreeding on immune response and body

size in a social insect, *Bombus terrestris. Funct Ecol,* 17, 582-9.
[http://dx.doi.org/10.1046/j.1365-2435.2003.00769.x]

Harbo, JR & Bolten, AB (1981) Development times of male and female eggs of the honey bee. *Ann Entomol Soc Am,* 74

Harrison, JM (1987) Roles of individual honeybee workers and drones in colonial thermogenesis. *J Exp Biol,* 129, 53-61.
[http://dx.doi.org/10.1242/jeb.129.1.53] [PMID: 3585245]

Harrison, JM (1986) Caste-specific changes in honeybee flight capacity. *Physiol Zool,* 59, 175-87.
[http://dx.doi.org/10.1086/physzool.59.2.30156031]

Haydak, MH (1957) The food of the drone larvae. *Ann Entomol Soc Am,* 50, 73-5. a
[http://dx.doi.org/10.1093/aesa/50.1.73]

Haydak, MH (1970) Honey Bee Nutrition. *Annu Rev Entomol,* 15, 143-56.
[http://dx.doi.org/10.1146/annurev.en.15.010170.001043]

Henderson, CE (1992) Variability in the size of emerging drones and of drone and worker eggs in honey bee (*Apis mellifera* L.) colonies. *J Apic Res,* 31, 114-8.
[http://dx.doi.org/10.1080/00218839.1992.11101271]

Hepburn, HR, Cantrill, RC, Thompson, PR & Kennedi, E (1979) Metabolism of carbohydrate, lipid and protein during development of sealed worker brood of the African honeybee. *J Apic Res,* 18, 30-5.
[http://dx.doi.org/10.1080/00218839.1979.11099939]

Hrassnigg, N, Brodschneider, R, Fleischmann, P & Crailsheim, K (2003) Worker bees (Apis mellifera L.) are able to utilize starch as fuel for flight while drones are not. *38th Apimondia Int Apic Congr*24–29 August 2003Ljubljana, Slovenia

Hrassnigg, N & Crailsheim, K (2005) Differences in drone and worker physiology in honeybees (*Apis mellifera*). *Apidologie (Celle),* 36, 255-77.
[http://dx.doi.org/10.1051/apido:2005015]

Woyke, J (1965) Genetic Proof of the Origin of Drones from Fertilized Eggs of the Honeybee. *J Apic Res,* 4, 7-11, 7-11.
[http://dx.doi.org/10.1080/00218839.1965.11100095]

Jay, SC (1963) The development of honeybees in their cells. *J Apic Res,* 2, 117-34.
[http://dx.doi.org/10.1080/00218839.1963.11100072]

Jaycox, ER (1961) The Effects of Various Foods and Temperatures on Sexual Maturity of the Drone Honey Bee (Apis mellifera)1. *Ann Entomol Soc Am,* 54, 519-23.
[http://dx.doi.org/10.1093/aesa/54.4.519]

Jones, JC, Helliwell, P, Beekman, M, Maleszka, R & Oldroyd, BP (2005) The effects of rearing temperature on developmental stability and learning and memory in the honey bee, Apis mellifera. *J Comp Physiol A Neuroethol Sens Neural Behav Physiol,* 191, 1121-9.
[http://dx.doi.org/10.1007/s00359-005-0035-z] [PMID: 16049697]

Jones, JC, Myerscough, MR, Graham, S & Oldroyd, BP (2004) Honey bee nest thermoregulation: diversity promotes stability. *Science,* 305, 402-4.https://doi. Org/10.1126/science.1096340
[http://dx.doi.org/10.1126/science.1096340] [PMID: 15218093]

Jones, JC & Oldroyd, BP (2006) Nest Thermoregulation in Social Insects.*Advances in Insect Physiology* Academic Press 153-91.
[http://dx.doi.org/10.1016/S0065-2806(06)33003-2]

Kairo, G, Provost, B, Tchamitchian, S, Ben Abdelkader, F, Bonnet, M, Cousin, M, Sénéchal, J, Benet, P, Kretzschmar, A, Belzunces, LP & Brunet, JL (2016) Drone exposure to the systemic insecticide Fipronil indirectly impairs queen reproductive potential. *Sci Rep,* 6, 31904.
[http://dx.doi.org/10.1038/srep31904] [PMID: 27549030]

Kaufman, BA, Kolesar, JE, Perlman, PS & Butow, RA (2003) A function for the mitochondrial chaperonin Hsp60 in the structure and transmission of mitochondrial DNA nucleoids in Saccharomyces cerevisiae. *J Cell Biol*, 163, 457-61.
[http://dx.doi.org/10.1083/jcb.200306132] [PMID: 14597775]

Kleinhenz, M, Bujok, B, Fuchs, S & Tautz, J (2003) Hot bees in empty broodnest cells: heating from within. *J Exp Biol*, 206, 4217-31.
[http://dx.doi.org/10.1242/jeb.00680] [PMID: 14581592]

Klenk, M, Koeniger, G, Koeniger, N & Fasold, H (2004) Proteins in spermathecal gland secretion and spermathecal fluid and the properties of a 29 kDa protein in queens of *Apis mellifera*. *Apidologie (Celle)*, 35, 371-81.
[http://dx.doi.org/10.1051/apido:2004029]

Knecht, D & Kaatz, HH (1990) Patterns of larval food production by hypopharyngeal glands in adult worker honey bees. *Apidologie (Celle)*, 21, 457-68.
[http://dx.doi.org/10.1051/apido:19900507]

Koeniger, G (1976) Einfluss der Kopulation auf den Beginn der Eiablage bei der Bienenkönigin (Apis mellifica L.). *Apidologie (Celle)*, 7, 343-55.
[http://dx.doi.org/10.1051/apido:19760405]

Koeniger, G (1981) In welchem Abschnitt des Paarungsverhaltens der Bienenkönigin findet die Induktion der Eiablage statt? *Apidologie (Celle)*, 12, 329-43.
[http://dx.doi.org/10.1051/apido:19810403]

Koeniger, G (1986) Mating sign and multiple mating in the honeybee. *Bee World*, 67, 141-50.
[http://dx.doi.org/10.1080/0005772X.1986.11098892]

Koeniger, G (1988) Mating flights of honey bee drones (Apis mellifera L.). Film documentation. *Biona Report*, 6, 29-34.

Koeniger, G, Koeniger, N, Mardan, M, Otis, G & Wongsiri, S (1991) Comparative anatomy of male genital organs in the genus Apis. *Apidologie (Celle)*, 22, 539-52.
[http://dx.doi.org/10.1051/apido:19910507]

Koeniger, N & Koeniger, G (1991) An evolutionary approach to mating behaviour and drone copulatory organs in Apis. *Apidologie (Celle)*, 22, 581-90.
[http://dx.doi.org/10.1051/apido:19910602]

Koeniger, N & Koeniger, G (2000) Reproductive isolation among species of the genus *Apis*. *Apidologie (Celle)*, 31, 313-39.
[http://dx.doi.org/10.1051/apido:2000125]

Koeniger, N, Koeniger, G, Gries, M & Tingek, S (2005) Drone competition at drone congregation areas in four *Apis* species. *Apidologie (Celle)*, 36, 211-21.
[http://dx.doi.org/10.1051/apido:2005011]

Koo, J, Son, TG, Kim, SY & Lee, KY (2015) Differential responses of Apis mellifera heat shock protein genes to heat shock, flower-thinning formulations, and imidacloprid. *J Asia Pac Entomol*, 18, 583-9.
[http://dx.doi.org/10.1016/j.aspen.2015.06.011]

Kovac, H & Stabentheiner, A (2002) Contribution of drones of different age to heat production in a honeybee colony. *Apidologie (Celle)*, 33, 500-1.

Kraus, FB, Neumann, P, van Praagh, J & Moritz, RFA (2004) Sperm limitation and the evolution of extreme polyandry in honeybees (Apis mellifera L.). *Behav Ecol Sociobiol*, 55, 494-501.
[http://dx.doi.org/10.1007/s00265-003-0706-0]

Kraus, B, Velthuis, HHW & Tingek, S (1998) Temperature profiles of the brood nests of *Apis cerana* and *Apis mellifera* colonies and their relation to varroosis. *J Apic Res*, 37, 175-81.
[http://dx.doi.org/10.1080/00218839.1998.11100969]

Kronenberg, F & Heller, HC (1982) Colonial thermoregulation in honey bees (Apis mellifera). *J Comp Physiol B,* 148, 65-76.
[http://dx.doi.org/10.1007/BF00688889]

Czekońska, K & Tofilski, A (2020) Body mass of honey bee drones developing in constant and in changing temperatures. *Apidologie (Celle),* 51, 510-8.
[http://dx.doi.org/10.1007/s13592-020-00738-5]

Kunert, K & Crailsheim, K (1987) Sugar and protein in the food for honeybee worker larvae.*Chemistry and biology of social insects, Verlag J* Peperny, München 164-5.

Lensky, Y, Cassier, P, Notkin, M, Delorme-Joulie, C & Levinsohn, M (1985) Pheromonal activity and fine structure of the mandibular glands of honeybee drones (Apis mellifera L.) (Insecta, Hymenoptera, Apidae). *J Insect Physiol,* 31, 265-76.
[http://dx.doi.org/10.1016/0022-1910(85)90002-2]

Lensky, Y & Rakover, Y (1983) Separate protein body compartments of the worker honeybee (Apis mellifera L.). *Comp Biochem Physiol,* 75B, 607-15.

Leonhard, B & Crailsheim, K (1999) Amino acids and osmolarity in honeybee drone haemolymph. *Amino Acids,* 17, 195-205.
[http://dx.doi.org/10.1007/BF01361882] [PMID: 10524277]

Levin, CG & Collison, CH (1990) Broodnest Temperature Differences and their Possible Effect on Drone Brood Production and Distribution in Honeybee Colonies. *J Apic Res,* 29, 35-44.
[http://dx.doi.org/10.1080/00218839.1990.11101195]

Li, Z, Huang, ZY & Sharma, DB (2016) Drone and Worker Brood Microclimates Are Regulated Differentially in Honey Bees, Apis mellifera. *PLOS ONE,* 11, e0148740.

Mackasmiel, LAM & Fell, RD (2000) Respiration rates in eggs of the honey bee, *Apis mellifera. J Apic Res,* 39, 125-35.
[http://dx.doi.org/10.1080/00218839.2000.11101032]

Matsuka, M, Watabe, N & Takeuchi, K (1973) Analysis of the food of larval drone honeybees. *J Apic Res,* 12, 3-7.
[http://dx.doi.org/10.1080/00218839.1973.11099724]

Mazeed, AM (2011) Morphometry and number of spermatozoa in drone honeybees (Hymenoptera: Apidae) reared under different conditions. *Eur J Entomol,* 108, 673-6.
[http://dx.doi.org/10.14411/eje.2011.085]

Melampy, RM, Willis, ER & McGregor, SE (1940) Biochemical aspects of the differentiation of the female honeybee (Apis mellifera L.). *Physiol Zool,* 13, 283-93.
[http://dx.doi.org/10.1086/physzool.13.3.30151576]

Metz, B & Tarpy, D (2019) Reproductive Senescence in Drones of the Honey Bee (Apis mellifera). *Insects,* 10, 11.
[http://dx.doi.org/10.3390/insects10010011] [PMID: 30626026]

(1900) Bau und Entwicklung des mannlichen Begattungsapparates der Honigbiene. Z.wiss. *Zoo,* 67, 439-60.

Mindt, B (1962) Untersuchungen über das Leben der Drohnen, insbesondere Ernährung und Geschlechtsreife. *Z Bienenforsch,* 6, 9-33.

Morse, RA, Strang, GE & Nowakowski, J (1967) Fall death rates of drone honey bees. *J Econ Entomol,* 60, 1198-202.
[http://dx.doi.org/10.1093/jee/60.5.1198]

Nelson, JA, Sturtevant, AP & Lineburg, B (1924) Growth and feeding of honeybee larvae, U.S. Dept. *Agric Bull No,* 1222, 1-38.

Hrassnigg, Norbert & Crailsheim, Karl (2005) Differences in drone and worker physiology in honeybees

(Apis mellifera). *Apidologie,* 36, 255-77.

O'Donnell, S & Beshers, SN (2004) The role of male disease susceptibility in the evolution of haplodiploid insect societies. *Proc R Soc Lond Ser B: Biol Sci,* 271, 979-383.
[http://dx.doi.org/10.1098/rspb.2004.2685]

Page, RE, Jr (1986) Sperm utilization in social insects. *Annu Rev Entomol,* 31, 297-320.
[http://dx.doi.org/10.1146/annurev.en.31.010186.001501]

Panzenböck, U & Crailsheim, K (1997) Glycogen in honeybee queens, workers and drones (Apis mellifera carnica Pollm.). *J Insect Physiol,* 43, 155-65.
[http://dx.doi.org/10.1016/S0022-1910(96)00079-0] [PMID: 12769919]

Radloff, SE, Hepburn, HR & Koeniger, G (2003) Comparison of flight design of Asian honeybee drones. *Apidologie (Celle),* 34, 353-8.
[http://dx.doi.org/10.1051/apido:2003031]

Rangel, J & Fisher, A, II (2019) Factors affecting the reproductive health of honey bee (Apis mellifera) drones—a review. *Apidologie (Celle),* 50, 759-78.
[http://dx.doi.org/10.1007/s13592-019-00684-x]

Reinhardt, E (1960) Kernverhältnisse, Eisystem und Entwicklungsweise von Drohnen- und Arbeiterinneneiern der Honigbiene (Apis mellifera). *Zool Jahrb Abt Anat Ontogenie Tiere,* 78, 167-234.

Retschnig, G, Williams, GR, Mehmann, MM, Yañez, O, de Miranda, JR & Neumann, P (2014) Sex-specific differences in pathogen susceptibility in honey bees (Apis mellifera). *PLoS One,* 9, e85261.
[http://dx.doi.org/10.1371/journal.pone.0085261] [PMID: 24465518]

Rhein, Wv (1951) Über die Ernährung der Drohnenmaden. *Z Bienenforsch,* 1, 63-6.

Currie, RW (1987) The Biology and Behaviour of Drones. *Bee World,* 68, 129-43.
[http://dx.doi.org/10.1080/0005772X.1987.11098922]

Rhodes, J (2002) Drone honey bees – rearing and maintenance.

Rortais, A, Arnold, G, Halm, MP & Touffet-Briens, F (2005) Modes of honeybees exposure to systemic insecticides: estimated amounts of contaminated pollen and nectar consumed by different categories of bees. *Apidologie (Celle),* 36, 71-83.
[http://dx.doi.org/10.1051/apido:2004071]

Rousseau, A & Giovenazzo, P (2016) Optimizing Drone Fertility With Spring Nutritional Supplements to Honey Bee (Hymenoptera: Apidae) Colonies. *J Econ Entomol,* 109, 1009-14.
[http://dx.doi.org/10.1093/jee/tow056] [PMID: 27018435]

Rowland, CM & McLellan, AR (1987) Seasonal changes of drone numbers in a colony of the honeybee, Apis mellifera. *Ecol Modell,* 37, 155-66.
[http://dx.doi.org/10.1016/0304-3800(87)90023-8]

Rueppell, O, Page, RE, Jr & Fondrk, MK (2006) Male behavioural maturation rate responds to selection on pollen hoarding in honeybees. *Anim Behav,* 71, 227-34.
[http://dx.doi.org/10.1016/j.anbehav.2005.05.008] [PMID: 18846249]

Ruiz-González, MX & Brown, MJF (2006) Males vs workers: testing the assumptions of the haploid susceptibility hypothesis in bumblebees. *Behav Ecol Sociobiol,* 60, 501-9.
[http://dx.doi.org/10.1007/s00265-006-0192-2]

Ruttner, H & Ruttner, F (1972) Untersuchungen über die Flugaktivität und das Paarungsverhalten der Drohnen. *Apidologie (Celle),* 3, 203-32.
[http://dx.doi.org/10.1051/apido:19720301]

Schlüns, H, Schlüns, EA, van Praagh, J & Moritz, RFA (2003) Sperm numbers in drone honeybees (*Apis mellifera*) depend on body size. *Apidologie (Celle),* 34, 577-84.
[http://dx.doi.org/10.1051/apido:2003051]

Schmickl, T & Crailsheim, K (2004) Inner nest homeostasis in a changing environment with special emphasis on honey bee brood nursing and pollen supply. *Apidologie (Celle)*, 35, 249-63.
[http://dx.doi.org/10.1051/apido:2004019]

Seeley, TD (1985) *Honeybee Ecology: A Study of Adaptation in Social Life.* Princeton University Pres.

Seeley, TD & Mikheyev, AS (2003) Reproductive decisions by honey bee colonies: tuning investment in male production in relation to success in energy acquisition. *Insectes Soc*, 50, 134-8.
[http://dx.doi.org/10.1007/s00040-003-0638-8]

Seeley, TD & Morse, RA (1976) The nest of the honey bee (Apis mellifera L.). *Insectes Soc*, 23, 495-512.
[http://dx.doi.org/10.1007/BF02223477]

Seidl, R (1980) Die Sehfelder und Ommatidien-Divergenzen der drei Kasten der Honigbiene (Apis mellifera). *Verh Dtsch Zool Ges*, 73, 367.

Simpson, J (1955) The significance of the presence of pollen in the food of worker larvae of the honeybee. *Q J Microsc Sci*, 96, 117-20.

Sinitzky, N & Lewtschenko, I (1971) Protein and free amino acid content in hemolymph of working honey bee individuals. *23rd Int Apic Apimondia Congr, Moskau*, 361-7.

Slone, JD, Stout, TL, Huang, ZY & Schneider, SS (2012) The influence of drone physical condition on the likelihood of receiving vibration signals from worker honey bees, Apis mellifera. *Insectes Soc*, 59, 101-7.
[http://dx.doi.org/10.1007/s00040-011-0195-5]

Smith, ML, Ostwald, MM & Seeley, TD (2015) Adaptive tuning of an extended phenotype: honeybees seasonally shift their honey storage to optimize male production. *Anim Behav*, 103, 29-33.
[http://dx.doi.org/10.1016/j.anbehav.2015.01.035]

Snodgrass, RE (1956) *Anatomy of the honey bee* Cornell University Press, London 334.

Stabe, HA (1930) The rate of growth of worker, drone and queen larvae of the honeybee, Apis mellifera Linn. *J Econ Entomol*, 23, 447-53.
[http://dx.doi.org/10.1093/jee/23.2.447]

Stabentheiner, A, Kovac, H & Brodschneider, R (2010) Honeybee Colony Thermoregulation – Regulatory Mechanisms and Contribution of Individuals in Dependence on Age, Location and Thermal Stress. *PLoS ONE*, 5, e8967.

Starks, PT, Blackie, CA & Seeley, TD (2000) Fever in honeybee colonies. *Naturwissenschaften*, 87, 229-31.
[http://dx.doi.org/10.1007/s001140050709] [PMID: 10883439]

Straus, J (1911) Die chemische Zusammensetzung der Arbeitsbienen und Drohnen während ihrer verschiedenen Entwicklungsstadien. *Z Biol (Münch)*, 56, 347-97.

Stürup, M, Baer-Imhoof, B, Nash, DR, Boomsma, JJ & Baer, B (2013) When every sperm counts: factors affecting male fertility in the honeybee Apis mellifera. *Behav Ecol*, 24, 1192-8.
[http://dx.doi.org/10.1093/beheco/art049]

Szentgyörgyi, H, Czekońska, K & Tofilski, A (2016) Influence of pollen deprivation on the fore wing asymmetry of honeybee workers and drones. *Apidologie (Celle)*, 47, 653-62.
[http://dx.doi.org/10.1007/s13592-015-0415-5]

Szentgyörgyi, H, Czekońska, K & Tofilski, A (2017) The Effects of Starvation of Honey Bee Larvae on Reproductive Quality and Wing Asymmetry of Honey Bee Drones. *J Apic Sci*, 61, 233-43.
[http://dx.doi.org/10.1515/jas-2017-0018]

Szentgyörgyi, H, Czekońska, K & Tofilski, A (2018) Honey bees are larger and live longer after developing at low temperature. *J Therm Biol*, 78, 219-26.
[http://dx.doi.org/10.1016/j.jtherbio.2018.09.007] [PMID: 30509639]

Szolderits, MJ & Crailsheim, K (1993) A comparison of pollen consumption and digestion in honeybee (Apis mellifera carnica) drones and workers. *J Insect Physiol*, 39, 877-81.

[http://dx.doi.org/10.1016/0022-1910(93)90120-G]

Taber, S & Roberts, WC (1963) Egg weight variability and its inheritance in the honey bee. *Ann Entomol Soc Am,* 56, 473-6.
[http://dx.doi.org/10.1093/aesa/56.4.473]

Taha, E-KA & Alqarni, AS (2013) Morphometric and Reproductive Organs Characters of Apis mellifera jemenitica Drones in Comparison to Apis mellifera carnica. *Int J Sci Eng Res,* 4.

Tan, Y, Zhang, Y, Huo, ZJ, Zhou, XR & Pang, BP (2018) Molecular cloning of heat shock protein 10 (Hsp10) and 60 (Hsp60) cDNAs from *Galeruca daurica* (Coleoptera: Chrysomelidae) and their expression analysis. *Bull Entomol Res,* 108, 510-22.
[http://dx.doi.org/10.1017/S0007485317001079] [PMID: 29081303]

Thrasyvoulou, AT & Benton, AW (1982) Rates of growth of honeybee larvae. *J Apic Res,* 21, 189-92.
[http://dx.doi.org/10.1080/00218839.1982.11100540]

de Oliveira Tozetto, S, Rachinsky, A & Engels, W (1995) Reactivation of juvenile hormone synthesis in adult drones of the honey bee,Apis mellifera carnica. *Experientia,* 51, 945-52.
[http://dx.doi.org/10.1007/BF01921745]

Trenczek, T, Zillikens, A & Engels, W (1989) Developmental patterns of vitellogenin haemolymph titre and rate of synthesis in adult drone honey bees (Apis mellifera). *J Insect Physiol,* 35, 475-81.
[http://dx.doi.org/10.1016/0022-1910(89)90054-1]

Vainio, L, Hakkarainen, H, Rantala, MJ & Sorvari, J (2004) Individual variation in immune function in the ant Formica exsecta; effects of the nest, body size and sex. *Evol Ecol,* 18, 75-84.
[http://dx.doi.org/10.1023/B:EVEC.0000017726.73906.b2]

Wang, DI (1965) Growth rates of young queen and worker honeybee larvae. *J Apic Res,* 4, 3-5.
[http://dx.doi.org/10.1080/00218839.1965.11100094]

Wharton, KE, Dyer, FC & Getty, T (2008) Male elimination in the honeybee. *Behav Ecol,* 19, 1075-9.
[http://dx.doi.org/10.1093/beheco/arn108]

Wharton, KE, Dyer, FC, Huang, ZY & Getty, T (2007) The honeybee queen influences the regulation of colony drone production. *Behav Ecol,* 18, 1092-9.
[http://dx.doi.org/10.1093/beheco/arm086]

aWharton, KE, Dyer, FC, Huang, ZY & Getty, T (2007) The honeybee queen influences the regulation of colony drone production. *Behav Ecol,* 18, 1092-9.
[http://dx.doi.org/10.1093/beheco/arm086]

bWharton, KE, Dyer, FC & Getty, T (2007) Male elimination in the honeybee. *Behav Ecol,* 19, 1075-9.
[http://dx.doi.org/10.1093/beheco/arm086]

Woyke, J (1998) Size change of *Apis mellifera* eggs during the incubation period. *J Apic Res,* 37, 239-46.
[http://dx.doi.org/10.1080/00218839.1998.11100978]

Woyke, J & Ruttner, F (1958) An anatomical study of the mating process in the honeybee. *Bee World,* 39, 3-18.
[http://dx.doi.org/10.1080/0005772X.1958.11095028]

Woyke, J (1963) Drones from fertilized eggs and the biology of sex-determination in the honey bee. *Bull Acad Po/on Sci Cl V,* 11, 251-4.

WoYKf, J. (1962) The hatchability of 'lethal' eggs in a two sex-allele fraternity of honeybees. *J Apic Res,* 6-13.

ZANDER, E. (1900) Beitrage zur Morphologie der mannlichen Geschlechtsanhange der Hymenopteren. Z.wiss. *Zoot,* 47, 461-89.

ZANDER, E., LoscHEL, F. & MEIER, K. (1916) Die Ausbildung des Geschlechtes bei der Honigbiene (Apis mellifica L.) Z. angew. *Ent,* 3, 1-74.

CHAPTER 4

The Pheromonal Profile of the Drone Honey Bees Apis mellifera: The Volatile Messengers

Abstract: The drone honey bee produces volatile chemicals during developmental and adult stages that facilitate chemical interaction between drone larvae and workers, drone to drone, and drone to the queen. For example, the drone larvae solicit larval food from nurse bees through chemical messengers; further, adult drones attract other drones to drone congregation areas (DCA) through pheromones; the drones attract queen honey bees to DCA through volatile chemicals. The mandibular drone gland secretes volatile chemicals, a mixture of saturated, unsaturated, and methyl-branched fatty acids. In the drone honey bee of *Apis mellifera*, about 18,600 olfactory poreplate sensilla per antenna are associated with receptor neurons. The current chapter highlights the role of chemical volatiles in Drone honey bee's life and overall influence.

Keywords: Chemical communication, Colonial cehaviour, Pheromones.

INTRODUCTION

The volatile chemicals and pheromones play a critical role in intra-specific communication in insect societies, including honey bees. Chemical communication governs interactions between larvae and adult workers, workers to workers, queens to queens, and queens to drones (Gary and Marston 1971; Pankiw *et al.* 1998; Slessor *et al.* 2005; Thom *et al.* 2007; Grozinger 2015). The honey bee colony is quite suitable in speculating the significance and complexity of pheromonal communication.

Due to their limited role in colonial productivity, the males remained relatively little in their exploration of chemical production and the involvement of specific chemicals in social communication. Therefore, few investigations are available that elaborate on the quantity, types, and function of the male sex-specific chemicals in the honey bee. Nevertheless, for a long time, drone honey bees have been assumed to play a role in colony resource consumption and reproduction (Boomsma and Ratnieks 1996; Tarpy and Nielsen 2002).

Lovleen Marwaha
All rights reserved-© 2023 Bentham Science Publishers

A few studies indicate that drones are engaged in various social interactions inside and outside the colony. According to Lensky *et al.* 1985; Boucher and Schneider 2009 studies, worker-drone and drone-drone interaction are facilitated by drone pheromone production. Further, the drones interact with the workers and the queen in the colony through chemical communication. As the drones get food from the nurse workers, they regularly engage in trophallaxis (Free 1957; Naumann *et al.* 1991).

Villar *et al.* (2018) reported that drones find and tightly aggregate with each other while inside a colony. The reason and implication of this specific aggregation are uncharacterized, but possibly drone-specific olfactory cues may support such an aggregation within a colony.

For mating, sexually mature drones fly out of the colony and aggregate at specific drone congregation areas (DCA) with other drones, waiting for the virgin queen to arrive (Ruttner and Ruttner 1972). The drone perceives and responds to queen pheromones 9-ODA (9-oxo2-decenoic acid) inside the colony, and specific chemicals further influence drone physiology and behaviour (Villar and Grozinger 2017). As the location of DCA is stable over time, it is probably due to the fact that the drone uses particular landscape features (GalindoCardona et al., 2015).

Further, according to Boucher and Schneider (2009), a specialized worker nursing stage preferentially targets sexually immature drones with a vibratory signal. Drones with specific signals solicit food from workers. The pathway for nurse bee particular behaviour for specific drone age groups is unknown but probably could be facilitated with a drone-produced, odour-mediated mechanism (Wakonigg *et al.* 2000).

THE PHEROMONAL COMMUNICATION IN THE COLONY

Pheromonal secretion regulates the colony's foraging, defensive, and brood-rearing activity (Free and Winder 1983; Free 1987; Vallet *et al.* 1991; Breed *et al.* 2004; Hunt 2007; Stout and Goulson 2001; Slessor *et al.* 2005; Maisonnasse *et al.* 2010). Slessor *et al.* 1990; Slessor *et al.* 2005; Grozinger 2015). The communication of the sex pheromones has been demonstrated by Gary (1962) in *Apis mellifera*. The queen's mandibular gland secretion includes 9-ODA (9-ket--2(E)-decenoic acid), due to which drones get attracted toward the queen. Further, other studies, including Butler *et al.* 1967; Sanasi *et al.* 1971, scrutinized other honey bee species, including *Apis florea, Apis cerana*, and *Apis dorsata* using the same type of pheromones.

Further, the communication between workers and developing larvae is regulated by pheromonal communications. Developing drone larvae are dependent upon the worker honey bees. Larvae release capping pheromones from their salivary gland containing ten methyl or ethyl fatty acid esters (Le Conte *et al.* 1989, 1990, 2006; Trouiller *et al.* 1991). The composition and quantity of pheromonal secretion vary by the age and sex of larvae. The workers adjust their behavioural responses to larvae accordingly (Free and Winder 1983; Le Conte 1994; Slessor *et al.* 2005). At the end of the final larval stage, workers secrete a thin wax cap over the cell to facilitate the pupation process within a clean and stable environment.

Le Conte *et al.* 1990 speculated that four components trigger the behaviour of workers, which include methyl palmitate (M.P.), methyl oleate (MO), methyl linoleate (ML), and methyl linolenate (MLN) (Le Conte *et al.* 1990). The pheromonal secretion reaches to peak when larvae are about to enter the pupation phase (Trouiller *et al.* 1991). Qin et al., 2019 analyzed the change in gene expression across the cell capping. Further, they analyzed the biosynthetic pathways for M.P., MO, ML, and MLN and a "housekeeping gene" (β-actin). For a specific analysis, larval samples had been taken from three different colonies and were stored in frozen liquid nitrogen at −80 °C until used. First, they analyzed the relative expression level by formula 2−ΔΔCt (Liu and Saint (2002), and after that, the square root was transformed to normalize the distribution of the data before ANOVA analysis. They found that M.P., MO, ML, and MLN acted as essential signals for honey bee capping behaviour. Similarly, Trouiller *et al.* (1992) concluded that drone honey bee larvae increase the amount of capping pheromones at the stage of wax capping. Consistently, Yan *et al.* (2009) also reported that in *Apis cerana* workers and drone larvae, there is a similar trend of hormonal release.

The drone cells are generally located at the edge of the bee hive, so Qin et al., 2019 took samples for study from the edges of the bee hive. Varroa mite is a global pest of honey bees (Le Conte *et al.* 2010). Varroa mites get attracted toward colonies with the influence of M.P., E.P., and ML (Le Conte *et al.* 1989). Acetyl-CoA is a precursor of different metabolic pathways which result in the synthesis of fatty acids and esters (Mcgee and Spector 1975; Cripps *et al.* 1986; Moshitzky *et al.* 2003). Insects can synthesize fatty acids de novo or get fatty acids in their diet. Yan *et al.* (2009) demonstrated through 13C and 2H isotope tracing that capping pheromone components were de novo synthesized by larvae. Furthermore, they have proposed that about 12 genes facilitate the processes of M.P., MO, ML, and MLN in honey bee larvae.

THE DRONE MANDIBULAR GLAND PHEROMONES

The mandibular drone glands secrete a blend of saturated, unsaturated, and methyl-branched fatty acids primarily, with chain lengths of nonanoic to docosanoic acids, which facilitate the chemical communication of drones with other drones, workers, and queens. Villar et al., 2017 elucidated the chemical composition of the extract of the mandibular drone gland and concluded that its blend of six main compounds facilitated social interaction in and out of the colony, whereas Lensky *et al.* 1985; and Brandstaetter *et al.* 2014, have scrutinized that the drone mandibular gland extract or group of five drones can attract drones within apiaries (Lensky *et al.* 1985; Brandstaetter *et al.* 2014), which highlights a chemically mediated mechanism for the formation of DCA. Drone mandibular pheromone biosynthesis occurs in adult life and ceases when the mating flight is initiated (Lensky *et al.* 1985). Therefore, it can be concluded that specific chemicals remain stored in the mandibular gland and are released during mating flights. Furthermore, they concluded that extracts of sexually mature drones differ in profile and function from immature males.

Villar *et al.* (2018) ascertained the role of mandibular gland pheromone in the drone's interaction with other drones and workers by testing the biological activity of DMG extracts and a synthetic blend inside and outside the hive. Furthermore, they identified 46 compounds, including saturated and unsaturated fatty acids, in the mandibular glands of drone honey bees. A natural extract of DMG recruited drones over long distances. Furthermore, they found a diverse assortment of compounds in the glands of 2-week-old, sexually mature drones taking mating flights. Their exploration concludes that DMG extract can attract other drones over a long distance, and they further concluded that extracts from sexually mature drones are biologically active.

Even Lensky *et al.* (1985) concluded that DMG extract could attract other drones within an apiary. Further studies concluded that drone-produced pheromones are partly responsible for forming and maintaining congregations (Gerig 1972; Lensky *et al.* 1985; Brandstaetter *et al.* 2014; Bastin *et al.* 2017). Such a response helps to perceive odour cues from other drones while taking nuptial flights or moving to nuptial flights and for movement. Earlier studies indicate that worker drone interaction is age-dependent (Boucher and Schneider 2009). Villar *et al.* (2018) concluded that workers respond similarly to the DMG extract and control. Drones' cuticular hydrocarbon profiles indicate the better drone age and status (Wakonigg *et al.* 2000). According to Boucher and Schneider (2009), the DMG compound cannot be specific to a specific nurse bee but rather to nurses who take care of the drones.

Villar *et al.* (2018) investigated a synthetic blend of six of the most abundant compounds and discovered a similar response in attracting drones from long and short distances. Long-term DMG responses suggest that drones must use volatiles to confirm their presence in DCA while waiting for a virgin queen. Here, the production of aggregation pheromones by drone could increase the competition but can also add an advantage if virgin queens use similar cues to locate DCA (Bastin *et al.* 2017). Compounds in DMG are present in a variety of insect taxa which are fatty acid-derived semiochemicals (Blum 1996). Some of the volatiles in DMG include hexadecanoic, 9-hexadecenoic, octadecanoic, and 9-octadecenoic acids in the mandibular glands of queen honey bees and even fluctuate with the queen's physiological state (Engels *et al.* 1997; Richard *et al.* 2007; Niño *et al.* 2013; Villar et al., 2018). According to Villar and Grozinger 2017, drones initiate flights at a 7-8 day old stage, whereas some other studies indicate that drones start mating flights at a 6-10 day old stage (Howell and Usinger 1956; Fukuda and Ohtani 1977; Rowell *et al.* 1986; Rueppell *et al.* 2006).

DETECTION OF PHEROMONES BY DRONE HONEY BEES

Drone honey bees attract the queen due to their pheromonal secretion (Fig. **1**). Male insects which get attracted to female mates over long distances possess better specialized olfactory systems. The specific sensory system consists of numerous receptor neurons sensitive to female pheromones. Further, olfactory receptors are sensitive to perceiving minute amounts of pheromones. In the *Apis mellifera*, there are about 18,600 olfactory poreplate sensilla per antenna, each associated with receptor neurons and further divided into four voluminous macroglomeruli (MG1–MG4) in the antennal lobes. In the case of *Apis florea*, each antenna possesses 1,200 poreplate sensilla per antenna and two macroglomeruli in their antennal lobes. In *Apis florea*, macroglomeruli are homologous, with the most prominent macroglomeruli in *Apis mellifera*, the MG1 and MG2, but comparatively slightly smaller. In *Apis florea*, fewer pheromones are required, whereas, in *Apis mellifera*, more complicated pheromonal processing is needed.

According to Brockmann and Brückner (2001), there are approximately 90 isomorphic glomeruli on the antennal lobes of *Apis florea* drones and about 100 glomeruli on the antennal lobes of *Apis mellifera* drones. The size of isomorphic glomeruli in *Apis mellifera* is double than that of *Apis florea* drones. Further, they identified two much larger glomeruli, a small one (44.3×103 μm3) in a frontal position and a larger one (77.3×103 μm3) in a dorsolateral position.

The queen's mandibular gland secretion is the main pheromone regulating queen–worker interaction (Free 1987). Slessor *et al.* 1988 demonstrated the behavioural significance of single components and composition differences that have been analyzed in different species of *Apis*. Kaissling and Renner (1968) reported that antennal poreplate sensilla in all morphs of *Apis mellifera* carry sensory neurons to 9-ODA. Drone honey bee antennae are more extended than worker honey bee antennae and contain seven times as many olfactory poreplate sensilla, with 18,600 in drones versus 2,600 in workers. *Apis mellifera* males and females exhibit sexual dimorphism in the glomerular organization of the antennal lobes. Drones of *Apis mellifera* possess a specific olfactory subsystem with receptor neurons on four macroglomeruli, whereas workers have fewer receptor neurons sensitive to mandibular pheromonal components.

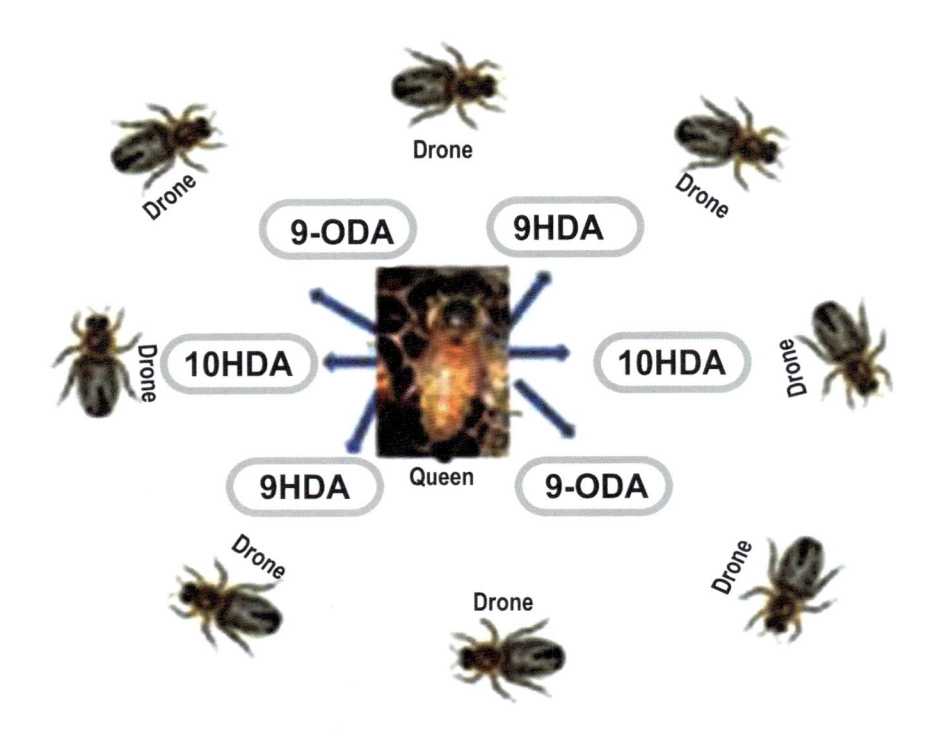

Fig. (1). Drone attraction to queen via queen pheromones such as 9-ODA, 9HDA, and 10 HDA.

DRONE CONGREGATION AREA

Loper et al., 1992, studied drone congregation areas (DCA's) in a nearby flat desert area with an X-band radar unit. They had screened an approximate 5.0 x 2.0 km area adjacent to a commercial apiary and had identified 26 DCA. They had considered DCA based on radar images with a diameter of 100 m, with a more significant number of drones. They discovered that the maximum height of drones in flyways was 21 meters, whereas they could fly from 30 to 50 metres above the ground in DCA.

In *Apis mellifera* L. Queen, honey bees mate outside the hive during nuptial flights. In a honey bee colony, more drone production occurs when there is more protein status in the colony (Weiss, 1969; Taber and Poole, 1974). Sexually mature drones fly in the afternoon, about an hour before flights of virgin queens. For mating, drones fly in drone congregation areas. In DCA, drones produce droning sounds due to the attraction of queen sex pheromones (9-oxodecenoic acid, 9-ODA. Elaborative information about DCA is not known, but certain constants include

Drones assemble at the exact location yearly, even without a queen.

- Drones responses are quite robust to queen pheromones in congregation areas.
- Pursuit of drones declines as queens move out of DCA.
- Drones flight altitude within DCA is inversely related to wind velocity.
- Drones quickly locate DCA.

Generally, drones fly to DCA, and after that, virgin females fly to DCA to mate. Drones fly for 2 or 3 km but can fly for 6 km or more. Weiss (1969) suggested that drones have a solidly joined system of orientation that directs them to a particular flight route. Tribe (1982) has described that environmental factors influencing DCA include wind, vertical relief, and thermal wind patterns. Loper et al., 1987 tried to study DCA with the help of a radar. They detected further flyways used by drones, DCAs near to apiary, and physical features of the site, which influence flyways or DCA and further determine the dynamics of drone flight. Loper et al., 1987; 1992, specified drone activity using X-band marine radar. They had determined

- The location, diameter, and height of drone activity in DCA.
- The width, height, and location of drone flyways.
- Drone's response to queen pheromones.

In *Apis* species, copulation during aerial flight is expected. During mating, drone honey bees maintain contact with the queen through the endophallus. In *Apis dorsata*, there is a particular adhesive organ on the hind legs of the drone. Further, in *Apis andreniformis* and *Apis florea*, drones have a reduced mass of mucus and therefore use their hind legs for copulation with the queen. The queen of *Apis mellifera* mates with about eight drones during the mating flight. A further queen can also mate within a second mating flight (Adams et al., 1977). After mating, the queen returns to the hive with approximately 120 million spermatozoa, of which about 7 million are stored in the spermatheca. In *Apis mellifera*, a large membranous endophallus connects the drone with the queen.

The Drone's Attraction Toward to Queen

Drones get attracted to virgin queens by getting attracted to chemicals produced by them. Therefore, they strongly follow the queen or any object impregnated with queen pheromones (Taylor, 2015).

In the case of *Apis mellifera*, the primary capping pheromones include methyl palmitate (M.P.), methyl oleate (MO), methyl linoleate (ML), and methyl linolenate (MLN). Qin et al., 2019 compared the capping pheromones in drones and worker larvae. In drone larvae, there are high capping pheromones throughout the development, and the level of capping pheromones further increases with ageing. Qin et al., 2019 proposed de novo biosynthetic pathways from acetyl-CoA for M.P., MO, ML, and ML.

Drones locate queen honey bees through anemotaxis. Drones orientate upward to detect volatile pheromones, which the queen produces. If the concentration of pheromones falls below a minimum threshold, in that case, drones fly randomly until they locate pheromones again, and after that, they continue upwind. In the presence of sex pheromones, drones get attracted to compact moving objects (Currie, 2016). Drones typically fly and become attracted to the queen at 10-50 m above ground. In a pre-mated condition, pheromones are secreted by her mandibular gland mainly. However, in 1962, Morse *et al.* found that a queen can still mate without a mandibular gland.

Some studies claim that the mandibular gland contains up to 32 compounds (Callow et al., 1964; Simpson, 1979). Two mandibular gland components attract drones, including 9-oxo-trans-2-decenoic acid (9-ODA) and 9-hydroxy-decenoic acid (9-HDA). It has been tested that 9-oxo-trans-2-decenoic acid (9-ODA) is more attractive than 9-HDA (Butler and Fairey, 1964). Even at higher concentrations, Blum et al., 1971; and Boch et al., 1975 found no drone attraction to 9-HDA. Pain and Ruttner (1963) reported that drones become more attracted to the queen mandibular gland than 9-ODA alone. Drones perceive information about pheromones through specific pore plate receptor sites on their antennae. In queen mandibular gland secretion, 9-oxo-2-decenoic acid is present in two isomeric forms. Trans isomers are about 200–400 times more attractive to drones than cis isomeric forms. Photoisomerization of the cis isomer to trans occurs under prolonged exposure to sunlight. The cis isomer possesses no masking effect on the activity of the trans isomer.

Fatty acid produced in the queen's head acts as a keeper substance to ensure the gradual release of pheromones. Drifting drones get attracted to a colony with a virgin queen but not with a mated queen. The virgin queen secretes a specific odour from glands in the abdominal tergites or sternites. Drone honey bees also produce certain chemicals that attract other drones, but specific glands degenerate in drones older than nine days (Currie, 1987).

CONCLUSION

Chemical messengers help in communication between different castes in the colony. This social interaction is required for the proper growth of larvae, food acquisition, mating and reproduction. Drones produce divergent types of chemical messengers that help develop and mate drones.

REFERENCES

Adams, J, Rothman, ED, Kerr, WE & Paulino, ZL (1977) Estimation of the number of sex alleles and queen matings from diploid male frequencies in a population of Apis mellifera. *Genetics,* 86, 583-96.
[http://dx.doi.org/10.1093/genetics/86.3.583] [PMID: 892423]

Axel Brockmann · Dorothea Brückner Structural differences in the drone olfactory system of two phylogenetically distant Apis species, A. florea and A. melliferous. *Naturwissenschaften,* 88, 78-81.
[http://dx.doi.org/10.1007/s001140000199] [PMID: 11320892]

Bastin, F, Cholé, H, Lafon, G & Sandoz, JC (2017) Virgin queen attraction toward males in honey bees. *Sci Rep,* 7, 6293.
[http://dx.doi.org/10.1038/s41598-017-06241-9] [PMID: 28740234]

Blum, MS (1996) Semiochemical parsimony in the Arthropoda. *Annu Rev Entomol,* 41, 353-74.
[http://dx.doi.org/10.1146/annurev.en.41.010196.002033] [PMID: 15012333]

(1971) Honeybee sex attractant: conformational analysis, structural specificity and lack of masking activity of congeners. *J Insect Physiol,* 7, 349-64.

Boch, R, Shearer, DA & Young, JC (1975) Honey bee pheromones: Field tests of natural and artificial queen substance. *J Chem Ecol,* 1, 133-48.
[http://dx.doi.org/10.1007/BF00987726]

Boomsma, JJ & Ratnieks, FLW (1996) Paternity in eusocial Hymenoptera. *Philos Trans R Soc Lond B Biol Sci,* 351, 947-75.
[http://dx.doi.org/10.1098/rstb.1996.0087]

Boucher, M & Schneider, SS (2009) Communication signals used in worker–drone interactions in the honeybee, *Apis mellifera. Anim Behav,* 78, 247-54.
[http://dx.doi.org/10.1016/j.anbehav.2009.04.019]

Brandstaetter, AS, Bastin, F & Sandoz, JC (2014) Honeybee drones are attracted by groups of consexuals in a walking simulator. *J Exp Biol,* 217, jeb.094292.
[http://dx.doi.org/10.1242/jeb.094292] [PMID: 24436379]

Breed, MD, Guzmán-Novoa, E & Hunt, GJ (2004) Defensive behavior of honey bees: organization, genetics, and comparisons with other bees. *Annu Rev Entomol,* 49, 271-98.
[http://dx.doi.org/10.1146/annurev.ento.49.061802.123155] [PMID: 14651465]

Butler, CG & Fairey, EM (1964) Pheromones of the honeybee: Biological studies of the mandibular gland secretion of the queen. *J Apic Res,* 3, 65-76.
[http://dx.doi.org/10.1080/00218839.1964.11100085]

Callow, RK, Chapman, JR & Paton, AN (1964) Pheromones of the honeybee: chemical studies of the mandibular gland secretion of the queen. *J Apic Res,* 3, 77-89.

[http://dx.doi.org/10.1080/00218839.1964.11100086]

Cripps, C, Blomquist, GJ & de Renobales, M (1986) De novo biosynthesis of linoleic acid in insects. *Biochim Biophys Acta Lipids Lipid Metab,* 876, 572-80.
[http://dx.doi.org/10.1016/0005-2760(86)90046-9]

Engels, W, Rosenkranz, P, Adler, A, Taghizadeh, T, Lübke, G & Francke, W (1997) Mandibular gland volatiles and their ontogenetic patterns in queen honey bees, *Apis mellifera* carnica. *J Insect Physiol,* 43, 307-13.
[http://dx.doi.org/10.1016/S0022-1910(96)00110-2] [PMID: 12769892]

Free, JB (1957) The food of adult drone honeybees (Apis mellifera). *Br J Anim Behav,* 5, 7-11.
[http://dx.doi.org/10.1016/S0950-5601(57)80038-0]

Free, JB (1987) *Pheromones of Social Bees.*Chapman & Hall, London.

Free, JB & Winder, ME (1983) Brood recognition by honeybee (*Apis mellifera*) workers. *Anim Behav,* 31, 539-45.
[http://dx.doi.org/10.1016/S0003-3472(83)80077-3]

Fukuda, H & Ohtani, T (1977) Survival and life span of drone honeybees. *Popul Ecol,* 19, 51-68.
[http://dx.doi.org/10.1007/BF02510939]

Villar, G, Wolfson, MD, Hefetz, A & Grozinger, CM (2018) Evaluating the Role of Drone-Produced Chemical Signals in Mediating Social Interactions in Honey Bees (Apis mellifera). *J Chem Ecol,* 44, 1-8.
[http://dx.doi.org/10.1007/s10886-017-0912-2]

Galindo-Cardona, A, Monmany, AC, Diaz, G & Giray, T (2015) A landscape analysis to understand the orientation of honey bee (Hymenoptera: Apidae) drones in Puerto Rico. *Environ Entomol,* 44, 1139.1-48.
[http://dx.doi.org/10.1093/ee/nvv099] [PMID: 26314058]

Gary, NE & Marston, J (1971) Mating behaviour of drone honey bees with queen models (Apis mellifera L.). *Anim Behav,* 19, 299-304.
[http://dx.doi.org/10.1016/S0003-3472(71)80010-6]

Gerald, M (1992) Loper, Wayne W. Wolf, and Orley R. Taylor, J Honey Bee Drone Flyways and Congregation Areas? *Radar Observation JOURNAL OF THE KANSAS ENTOMOLOGICAL SOCIETY,* 65, 223-30.

Gerig, L (1972) Ein weiterer Duftstoff zur Anlockung der Drohnen von Apis mellifica (L.)1. *Z Angew Entomol,* 70, 286-9.
[http://dx.doi.org/10.1111/j.1439-0418.1972.tb02183.x]

Grozinger, CM (2015) *The hive and the honey bee.*Dadant, Indianapolis.

Oertel, E & Usinger, RL (1956) Observations on the flight and length of life of drone bees. *Ann Entomol Soc Am,* 49, 497-500.
[http://dx.doi.org/10.1093/aesa/49.5.497]

Hunt, GJ (2007) Flight and fight: A comparative view of the neurophysiology and genetics of honey bee defensive behavior. *J Insect Physiol,* 53, 399-410.
[http://dx.doi.org/10.1016/j.jinsphys.2007.01.010] [PMID: 17379239]

Conte, YL, Bécard, JM, Costagliola, G, de Vaublanc, G, Maâtaoui, ME, Crauser, D, Plettner, E & Slessor, KN (2006) Larval salivary glands are a source of primer and releaser pheromone in honey bee (Apis mellifera L.). *Naturwissenschaften,* 93, 237-41.
[http://dx.doi.org/10.1007/s00114-006-0089-y] [PMID: 16541233]

Le Conte, Y, Sreng, L & Trouiller, J (1994) The recognition of larvae by worker honeybees. *Naturwissenschaften,* 81, 462-5.
[http://dx.doi.org/10.1007/BF01136651]

Le Conte, Y, Arnold, G, Trouiller, J, Masson, C, Chappe, B & Ourisson, G (1989) Attraction of the parasitic mite varroa to the drone larvae of honey bees by simple aliphatic esters. *Science,* 245, 638-9.

[http://dx.doi.org/10.1126/science.245.4918.638] [PMID: 17837619]

Le Conte, Y, Arnold, G, Trouiller, J, Masson, C & Chappe, B (1990) Identification of a brood pheromone in honeybees. *Naturwissenschaften,* 77, 334-6.
[http://dx.doi.org/10.1007/BF01138390]

Lensky, Y, Cassier, P, Notkin, M, Delorme-Joulie, C & Levinsohn, M (1985) Pheromonal activity and fine structure of the mandibular glands of honeybee drones (Apis mellifera L.) (Insecta, Hymenoptera, Apidae). *J Insect Physiol,* 31, 265-76.
[http://dx.doi.org/10.1016/0022-1910(85)90002-2]

Liu, W & Saint, DA (2002) A new quantitative method of real time reverse transcription polymerase chain reaction assay based on simulation of polymerase chain reaction kinetics. *Anal Biochem,* 302, 52-9.
[http://dx.doi.org/10.1006/abio.2001.5530] [PMID: 11846375]

Maisonnasse, A, Lenoir, JC, Beslay, D, Crauser, D & Le Conte, Y (2010) E-β-ocimene, a volatile brood pheromone involved in social regulation in the honey bee colony (*Apis mellifera*). *PLoS One,* 5, e13531.
[http://dx.doi.org/10.1371/journal.pone.0013531] [PMID: 21042405]

McGee, R & Spector, AA (1975) Fatty acid biosynthesis in Erlich cells. The mechanism of short term control by exogenous free fatty acids. *J Biol Chem,* 250, 5419-25.
[http://dx.doi.org/10.1016/S0021-9258(19)41198-8] [PMID: 237919]

Morse, RA, Strang, GE & Nowakowski, J (1967) Fall death rates of drone honeybees. *J Econ Entomol,* 60, 1198-202.
[http://dx.doi.org/10.1093/jee/60.5.1198]

Moshitzky, P, Miloslavski, I, Aizenshtat, Z & Applebaum, SW (2003) Methyl palmitate: a novel product of the Medfly (Ceratitis capitata) corpus allatum. *Insect Biochem Mol Biol,* 33, 1299-306.
[http://dx.doi.org/10.1016/j.ibmb.2003.06.008] [PMID: 14599501]

Naumann, K, Winston, ML, Slessor, KN, Prestwich, GD & Webster, FX (1991) Production and transmission of honey bee queen (*Apis mellifera* L.) mandibular gland pheromone. *Behav Ecol Sociobiol,* 29, 321-32.
[http://dx.doi.org/10.1007/BF00165956]

Niño, EL, Malka, O, Hefetz, A, Tarpy, DR & Grozinger, CM (2013) Chemical profiles of two pheromone glands are differentially regulated by distinct mating factors in honey bee queens (*Apis mellifera* L.). *PLoS One,* 8, e78637.
[http://dx.doi.org/10.1371/journal.pone.0078637] [PMID: 24236028]

Taylor, OR, Jr (1984) An Aerial Trap for Collecting Drone Honeybees in Congregation Areas. *J Apic Res,* 23, 18-20.
[http://dx.doi.org/10.1080/00218839.1984.11100603]

Pankiw, T, Huang, Z, Winston, ML & Robinson, GE (1998) Queen mandibular gland pheromone influences worker honey bee (*Apis mellifera* L.) foraging ontogeny and juvenile hormone titers. *J Insect Physiol,* 44, 685-92. a
[http://dx.doi.org/10.1016/S0022-1910(98)00040-7] [PMID: 12769952]

Pankiw, T, Page, RE, Jr & Kim Fondrk, M (1998) Brood pheromone stimulates pollen foraging in honey bees (Apis mellifera). *Behav Ecol Sociobiol,* 44, 193-8.
[http://dx.doi.org/10.1007/s002650050531]

Qin, Q-H, He, X-J, Barron, AB, Guo, L & Jiang, WJ The capping pheromones and putative biosynthetic pathways in worker and drone larvae of honey bees Apis mellifer. *Apidologie.*

Richard, FJ, Tarpy, DR & Grozinger, CM (2007) Effects of insemination quantity on honey bee queen physiology. *PLoS One,* 2, e980.
[http://dx.doi.org/10.1371/journal.pone.0000980] [PMID: 17912357]

Currie, RW (1987) The Biology and Behaviour of Drones. *Bee World,* 68, 129-43.
[http://dx.doi.org/10.1080/0005772X.1987.11098922]

Rowell, GA, Taylor, OR, Jr & Locke, SJ (1986) Variation in drone mating flight times among commercial honey bee stocks. *Apidologie (Celle)*, 17, 137-58.
[http://dx.doi.org/10.1051/apido:19860206]

Rueppell, O, Page, RE, Jr & Fondrk, MK (2006) Male behavioural maturation rate responds to selection on pollen hoarding in honeybees. *Anim Behav*, 71, 227-34.
[http://dx.doi.org/10.1016/j.anbehav.2005.05.008] [PMID: 18846249]

Ruttner, H & Ruttner, F (1972) Investigations on the flight activity and mating behaviour of drones: drone congregation areas and mating distance. *Apdologie*, 3, 203-32.
[http://dx.doi.org/10.1051/apido:19720301]

Simpson, J (1979) The existence and physical properties of pheromones by which worker honeybees recognize queens. *J Apic Res*, 18, 233-49.
[http://dx.doi.org/10.1080/00218839.1979.11099976]

Slessor, KN, Kaminski, LA, King, GGS & Winston, ML (1990) Semiochemicals of the honeybee queen mandibular glands. *J Chem Ecol*, 16, 851-60.
[http://dx.doi.org/10.1007/BF01016495] [PMID: 24263600]

Slessor, KN, Winston, ML & Le Conte, Y (2005) Pheromone communication in the honeybee (*Apis mellifera* L.). *J Chem Ecol*, 31, 2731-45.
[http://dx.doi.org/10.1007/s10886-005-7623-9] [PMID: 16273438]

Stout, JC & Goulson, D (2001) The use of conspecific and interspecific scent marks by foraging bumblebees and honeybees. *Anim Behav*, 62, 183-9.
[http://dx.doi.org/10.1006/anbe.2001.1729]

Tarpy, DR & Nielsen, DI (2002) Sampling error, effective paternity, and estimating the genetic structure of honey bee colonies (Hymenoptera: Apidae). *Ann Entomol Soc Am*, 95, 513-28.
[http://dx.doi.org/10.1603/0013-8746(2002)095[0513:SEEPAE]2.0.CO;2]

Thom, C, Gilley, DC, Hooper, J & Esch, HE (2007) The scent of the waggle dance. *PLoS Biol*, 5

Trouiller, J, Arnold, G, Le Conte, Y, Masson, C & Chappe, B (1991) Temporal pheromonal and kairomonal secretion in the brood of honeybees. *Naturwissenschaften*, 78, 368-70.
[http://dx.doi.org/10.1007/BF01131612]

Vallet, A, Cassier, P & Lensky, Y (1991) Ontogeny of the fine structure of the mandibular glands of the honeybee (*Apis mellifera* L.) workers and the pheromonal activity of 2-heptanone. *J Insect Physiol*, 37, 789-804.
[http://dx.doi.org/10.1016/0022-1910(91)90076-C]

Villar, G & Grozinger, CM (2017) Primer effects of the honeybee, *Apis mellifera*, queen pheromone 9-ODA on drones. *Anim Behav*, 127, 271-9.
[http://dx.doi.org/10.1016/j.anbehav.2017.03.023]

Wakonigg, G, Eveleigh, L, Arnold, G & Crailsheim, K (2000) Cuticular hydrocarbon profiles reveal age-related changes in honey bee drones (*Apis mellifera carnica*). *J Apic Res*, 39, 137-41.
[http://dx.doi.org/10.1080/00218839.2000.11101033]

CHAPTER 5

The Mating and Reproduction in Apis Mellifera: The Role of Drone Honey Bee

Abstract: Mating in honey bees occurs in the drone congeration area, where the queen and drones gather for reproduction. Sexually mature drones from all colonies congregate there, and the queen mates with multiple drones, increasing patrilinear inheritance variation. The variation in genomic content results in specific colonial behaviour, productivity, and strength. The current chapter discusses mating in honey bee colonies.

Keywords: Haploid drones, Diploid drones, Pheromones, Reproductive system.

INTRODUCTION

Apis mellifera colony consists of a polyandrous queen, several thousand facultative sterile female workers, and a few thousand seasonal males (drones), constitute a colony. Multiple mating increases genetic variability within a colony, which may have a selective value. Such colonies might cope better with changing environmental conditions (Crozier and Page 1985; Robinson and Page 1988). Polyandry further reduces the chances of parasites or pathogens destroying the colony to the extent of survival and reproduction (Sherman *et al.* 1988). In a colony with several sex alleles in a population, there is a high brood survival rate (Page 1980; Page and Metcalf 1982).

Further, polyandry conditions have resulted in complex flight behaviour. Polyandry in honey bees is not caused by insufficient sperm production by drones; a single drone produces more sperms than the spermatheca can hold. Instead, polyandry in honey bees occurs due to some other advantages like genetic variation.

Mating of queen and drone occurs in the drone congregation area, a location away from the hive, further increasing flight time and associated survival risk. During the nuptial flight, drones compete with one another, and the fastest drone will be able to mate. The mechanism for the queen's preference for her mate is unknown

Lovleen Marwaha
All rights reserved-© 2023 Bentham Science Publishers

(Kerr and Bueno 1970). The drones, after mating, support the succeeding drones to mate with the queen. As such, a drone marks the queen with conspicuous colour to reduce the mating-flight time (Koeniger, 1990).

In a colony, drone rearing occurs during the reproductive season when resources are plentiful and a large workforce is ready for rearing (Winston 1987; Rowland and McLellan 1987; Rangel *et al.* 2013; Rangel *et al.* 2013; Smith *et al.* 2014). As a result, male production occurs with the construction of comb cells with comparatively larger sizes than worker-destined cells (Seeley and Morse 1976; Boes 2010; Smith *et al.* 2014). Further, there is a variation in drone cells, which further results in variation in the adults' size (Berg 1991; Berg *et al.* 1997; Schlüns *et al.* 2003; Berg *et al.* 1997; Couvillon *et al.* 2010).

Generally, drones are produced about 3–4 weeks before the production of a new queen so that a new queen formed during the reproductive season would be able to get sexually mature drones for reproduction (Page 1981). Drones mate once, but the queen exhibits extreme polyandry by mating with 12 to 14 drones, which can go up to 50 or more drones (Estoup 1995; Tarpy and Page 2000; Rhodes 2002; Abdelkader *et al.* 2014; Palmer and Oldroyd 2000; Koeniger *et al.* 2005a; Amiri *et al.* 2017; Brutscher *et al.* 2019). After a few days of hatching, the young queen honey takes several nuptial flights to drone congregation areas, where she mates in the air with an average of 12 drones from other colonies (Tarpy and Nielsen, 2002). The exact number and origin of drones are difficult to observe (Koeniger et al., 2015). Artificial insemination has been developed as a practical tool for economic breeding (Laidlaw, 1987; Nolan, 1932).

Drone development needs 24 days, with queen development 16 days and 21 days. Further, this developmental time can vary depending upon factors like haplotype, temperature, and overall colony conditions (DeGrandiHoffman *et al.* 1998; DeGrandiHoffman 1993; Biekowska *et al.* 2011; Sturup *et al.* 2013; Winston 1987; DeGrandi-Hoffman 1993; Collison 2004). Few reports witness that in the case of honey bees, spermatogenesis starts at the larval stage and is completed at the pupal stage (Holldobler and Bartz 1985; Bishop 1920; Hoage and Kessel 1968). A drone ejaculates between 0.91 and 1.7 µL per drone, containing about 3.6 to 12 million sperm cells (Woyke 1960; Nguyen 1995; Collins and Pettis 2001; Rhodes 2008; Rousseau *et al.* 2015; Mackensen 1955; Woyke 1962; Duay *et al.* 2002; Schlüns *et al.* 2003; Rhodes *et al.* 2011). Further, sperm count is influenced by size, larval diet, and season (Nguyen 1995; Schlüns *et al.* 2003; Rhodes 2011).

During mating, the drone honey bee congregates in specific areas independently of the presence of the queen (Jean-Prost 1958; Ruttner & Ruttner 1963, 1965;

Zmarlicki & Morse 1963). After the entry of a queen honey bee in that area, drones just pursue her like a comet-like group due to queen sex pheromones and visual cues (Gary 1962, 1963; Pain & Ruttner 1963; Strang 1970; Gerig 1971).

For DCA (**Drone Congregation Area**), about 11,000 drones gather in mid-air, about 10 to 50 m above ground (Free 1987; Baudry et al. 1998; Koeniger et al. 2005a). Drones emit specific volatiles, which modulate their social interaction and further assist in forming DCAs (Villar et al., 2018; Brandstaetter et al., 2014). When virgin queens enter the congregation area, they attract drones with pheromones, in particular, 9-oxo-2-decenoic acid, or with visual cues at short range, which focus on finding and matching with the queen (9-ODA; Brandstaetter et al. 2014; Gries and Koeniger 1996; Baudry et al. 1998; Jaffé and Moritz 2010; Goins and Schneider 2013).

Virgin queens visit DCA on one or several mating flights, which can happen in one or several days (Roberts 1944; Tarpy and Page 2000). DCA comprises drones from colonies located 5 km away from each other (Free 1987; Baudry et al. 1998; Koeniger et al. 2005a). The drone hone bee gathers away from the colony to avoid inbreeding (Winston 1987). Factors that promote drones' gathering to DCA are poorly understood (Koeniger et al. 2005b).

Mating in honey bee colonies occurs in the drone congregation area. Further, properly exploring these areas is very important for keepers to get queen mating. Abou-Shaaraa and Kelany, 2021) explored DCA using remote sensing and geographical information systems. Mating of honey bee queens occurs in the air with numerous drones (Cobey, 2007; Moritz et al., 1996; Neumann & Moritz, 2000). These areas possess specific characteristics which is why the same site is chosen repeatedly (Ruttner & Ruttner, 1972).

During mating, at the beginning of the eversion of the membranous endophallus, drones become paralyzed. With the queen's movements, eversion of the endophallus proceeds, and sperm injection occurs in the oviduct. After mating, drones drop to the ground and die. The mating signs include mucus from male accessory glands, expelled chitin plates from drone endophallus, and a sticky orange layer from cornual gland cells (Koeniger 1988). The honey bee colony size is positively proportional to the number of queens produced for reproduction and the number of swarms (Winston 1979; Winston & Taylor 1980).

DRONE CONGREGATION AREA

Drone honey bees and queens gather in the Drone Congregation Areas (DCAs). DCA is formed in places with characteristic features. Galindo-Cardona, 2012, studied DCA with the help of a helium balloon equipped with queen-se-

-pheromone-impregnated bait and determined the presence of a high number of drones. Identification of diagnostic features of drone congregation areas (DCAs) is quite crucial for a better understanding of bee behaviour and for quality control of honey bee colonies for genetically programmed DCA in a location outside the colony of *Apis mellifera* L., where hundreds of drones and a queen assemble in the afternoon hours. Mating occurs during flight, usually at a height of 15 to 50 m, with bees flying at 12 kilometres per hour (Oertel 1956; Zmarlicki and Morse 1963; Loper *et al.* 1992). However, all matings do not occur at DCA, as some matings occur in the flight path of drones (Koeniger *et al.* 2005).

In the drone congregation area, queens are encountered by more than 1000 drones, with extreme competition and a low probability of mating. The drones form a cluster of about 20–100 near one queen. A drone mating with a queen attracts other drones toward the queen. Post mating, the drone leaves a mating sign in the queen, which facilitates further mating of other drones. Additionally, it keeps the queen sting chamber open and protects the endophallus of other drones against the queen's sting (Koeniger 1986).

DCA is formed at a height of 15 to 50 m at a distance of 500 m to 5 km (Ruttner, 1976). The drones prefer the shortest path for rapid mating (Gries and Koeniger, 1996; Koeniger et al., 2005). DCA is selected away from roads and dwellings, usually in exposed spaces (Zmarlicki and Morse, 1963; Galindo-Cardona et al., 2012). Trees shield DCA from the wind (Scheiner et al., 2013). Therefore, sites with trees in open areas are suitable for DCA. Many factors hinder queens' mating, including predators, air temperatures, cloud cover, and wind (Oertel, 1956; Verbeek, 1976; Lensky and Demter, 1985). The bee hive can be placed near DCA to ensure the successful mating of virgin queens. Loper et al., 1987 used radar for the location of the DCA (Abou-Shaaraa and Kelany, 2021).

The honey bee colonies are faithful to the DCA as these areas are visited by different generations of drones and queens (Laidlaw and Page 1984; Schlüns *et al.* 2005). According to one hypothesis, it is because of the specific physical characteristics of the areas (Winston 1987). One factor which is negatively proportional to drones' visits to DCA is the distance of the mating site from the apiary (Koeniger *et al.* 2005). Another hypothesis, the behavioural DCA hypothesis, points out that DCA results from the behavioural interaction of flying drones with a queen (Loper *et al.* 1992). Studying DCA is essential for understanding animal navigation, conservation, and population genetics. A few DCA explorations facilitate the estimation of genetic diversity and genetic structure (Evans 2006; Robinson *et al.* 2008;Collet et al., 2009). Another hypothesis says that drones detect features of the landscape better within a similar range when drones fly at or around the DCA (Galindo-Cardona et al.,2012).

DRONE FLIGHTS BEFORE MATING

Young drones interact with workers near the brood area to be fed and groomed (Goins and Schneider 2013; Collison 2004). After 5 to 8 days of emergence, drones take orientation flights to learn about the local landmarks and precise locations (Tofilski and Kopel 1996; Collison 2004; Galindo-Cardona *et al.* 2015). Sexually mature drones, after learning landmarks and the location of the hive, join the drone congregation area with a diameter of about 30 to 200 metres (Loper *et al.* 1987, 1992; Koeniger and Koeniger 2004).

A few factors affect drone flight navigation, including vegetation structure, directionality, and density (Galindo-Cardona *et al.* 2012, 2015). Slone *et al.* (2012) demonstrated that workers produce more trophallaxis stimulating vibration signals toward drones with poor flying capability, which probably encourages them to be more competitive. After copulation, the slender end of the endophallus breaks off and is pushed into the sting chamber of the queen's reproductive tract (Koeniger 1990; Woyke 2008). Successful mating is fatal for drones as their endophallus remains lodged in the genital tract of the queen (Page 1986; Woyke 2008; Goins and Schneider 2013). The queen with an endophallus as a mating sign attracts more drones. This further highlights cooperative communication in queens with other drones (Koeniger 1990). Unmated drones typically possess a life span of 20–40 days after emergence and are evicted from the hive by workers (Page and Peng 2001; Stürup *et al.* 2013; Metz and Tarpy 2019; Rhodes 2002). Eviction of drones from a colony is influenced by temperature, time of year, and food source (Rhodes 2002). Drone eviction generally occurs at the end of the reproductive season when drones are no longer needed (Winston 1987). The drones do not have elongated proboscis and corbiculae like workers, which are required for foraging (Hrassnigg and Crailsheim 2005). The drones help in passive colony thermal regulation through their collective presence as a part of a tight cluster (Fahrenholz *et al.* 1992).

THE DRONE'S REPRODUCTIVE POTENTIAL

An *Apis mellifera* colony is a composite unit of one queen, thousands of females, and a few hundred drones, which reproduce when colony resources are plentiful. An essential function of the drone is to mate with a virgin queen, which helps transfer their colony's genes to their mates. Several environmental and in-hive conditions influence the quality and viability of drones. Rangel and Fisher (2019) studied how different factors may influence drones' reproductive health, including nutrition, temperature, season, and age. Sperm development occurs during pupation in drones. Therefore, factors that influence drone development could also impact gamete production. Eusocial Hymenopteran species possess haploid-diploid sex determination systems, which comprise the formation of a male from an unfertilized egg and a female from a diploid egg (Wilson 1971; Palmer and Oldroyd 2000; Collison 2004). In honey bee colonies, males are nurtured in their colony by sister workers until they reach sexual maturity (Stürup et al., 2013). Male honey bees are not involved in any colony function therefore challenging to maintain (Holldobler and Bartz 1985).

The drone's reproductive potential is influenced by age, season, and genetics. Various studies have shown that drone senescence reduces sperm viscosity and viability (Woyke and Jasiski 1978; Cobey 2007; Czekoska et al. 2013a; Woyke and Jasiski 1978; Locke and Peng 1993; Rhodes et al. 2011; Czekoska et al. 2013a; Stürup et al. 2013), whereas few other studies show constant or decreasing sperm viability over time (Metz and Tarpy 2019; Czekońska et al. 2013a; Woyke and Jasiński 1978; Cobey 2007). Drones older than 21 days have thicker sperm, which makes it difficult for the queen to expel to the oviduct (Woyke and Jasiski 1978; Czekoska et al. 2013a). An investigation by Locke and Peng (1993) indicated that ageing affects sperm viability, which decreased from 86% to 80% when drones reached the age of 14–20 days old. Sturup et al. (2013) reported that drones with more than 20 days of generation possess 50% lower sperm viability. Although the impact of ageing on drone reproductive quality is variable, Metz and Tarpy (2019) discovered that sperm viability in drones remains constant throughout their lives. Czekońska et al. (2013a) show that semen volume and sperm viability increase with age. Metz and Tarpy (2019) reported that sperm count in drone semen increases to a peak at around 20 days post-emergence. Rhodes et al. (2011) explored the influence of age and season on drone semen volume and sperm counts. Rousseau et al. (2015) investigated the effects of season and age on drones aged 21 days and 14-35 days. Zaitoun et al. (2009) found that the male *Apis mellifera* ligustica produced more sperm than the Syrian bee *Apis mellifera* syriaca.

After the first week of emergence, migration of sperm takes place to the seminal vesicle along with a pair of mucus glands for protection and nourishment of sperm (Snodgrass 1956; Woyke 1983; Rhodes 2008; Johnson et al. 2013; Rousseau et al. 2015). Sperm cells receive protection from pathogens through proteins contained in seminal fluid (Peng et al. 2016). Such a protein composition is required for sperm viability and longevity (Baer et al., 2009; King et al., 2011). Some factors, including climate, nutrition, and other environmental factors, influence sexual maturation (Rhodes 2008). Drones generally reach sexual maturity by 6 to 8 days (Bishop, 1920; Mackensen and Roberts, 1948), from 10 to 12 days (Woyke and Ruttner, 1958; Moritz, 1989; Nguyen, 1995), or even up to 16 days (Rhodes 2002).

As this secretion fills the queen's sting chamber and the queen's bursa copulatrix, it is considered to function as a mating plug, which can prevent or complicate subsequent matings (Thornhill & Alcock 1983). However, the next drone can easily remove the mating sign with the help of a particular set of hair on the endophallus. Vision is essential for mate recognition in honey bees. An average colony rears between 5000 and 20,000 drones and about ten queens per year (Weiss 1962; Allen 1963, 1965; Gary 1963; Ruttner 1966;Strang 1970;Winston 1980; Seeley 1985).

Further, the genetic makeup of the colony influences the drone quality. Drones produced by worker honey bees are comparatively smaller in size and produce fewer normal sperm than the drones produced by queen honey bees (Gençer and Firatli 2005; Zaitoun *et al.* 2009). In a colony, different genetic lines have drones of varying body weights, wing morphology, and sperm counts. For example, Taha and Alqarni (2013) reported that drones of *Apis mellifera* carnica are heavier and produce more sperm than drones of *Apis mellifera* jemenitica. According to Rhodes *et al.* (2011), the number of sperm and ejaculatory volumes depends upon the colony's genetic line. Drones' quality depends on genetic constitution, nutrition, location, age, and season (Rangel and Fisher, 2019).

The Queen honey bee mates with an average of 12 drones and receive about 6 million spermatozoa into its oviduct from each male. About 5.5 million sperm are transported to the spermatheca by active and passive mechanisms over 40 hours (Laidlaw and Page 1984). Post-mating, there is a change in behaviour, physiology, and interaction with workers. Poorly inseminated queens produce different mandibular gland pheromones and are less attractive to workers.

Semen quality influences mating success in honey bees. Spring-reared drones possess a higher volume of semen than summer or autumn-reared drones. It has been concluded that sperms develop in the testes of drones during their pupal stage. Moritz (1989) described the reproductive organs and mating procedure of drones. Reproductive organs include the endophallus (copulatory organ), testes, seminal vesicles, and mucous glands. Endophallus is a long duct within the abdomen with three distinct zones–the vestibulum, the cervix, and the bulb. Migration of sperm from testes to seminal vesicles begins when drones are only 2-3 days old. Developing sperms attach their heads with gland cells, and the secondary physiological process of maturation begins.

Furthermore, gland cells empty their contents into vesicles among spermatozoa. Mucous secretion begins after emergence and is completed by day 5. A drone sperm consists of two components: sperm from the testis and seminal fluid from the seminal vesicles. Bishop (1920) reported that drones mature in 5–6 days.

Moritz (1989) stated that drones of twelve days are mature and can be used for insemination of queen bees. Van Niem Nguyen (1995) reported that drones fed on improved protein nutrition reach sexual maturity as early as ten days. Workers generally destroy diploid drone larvae. Colonies placed 2.5 km from DCA have better chances of mating.

There are few reports on drone fertility and potential impact because drones are only common during the summer. According to Rhodes (2002), drones mature at about 16 days and become less effective for mating at 28 days, whereas according to Moritz (1989), drones mature by about 12 days of age. The time required for sexual maturity depends upon the drone's genetics (Rhodes et al., 2010). Other factors which influence the maturation of drones include sources of pollen and diet (Cobey 1983; Mackensen and Roberts, 1948). Abdelkader *et al.* (2013) studied the sexual maturity of drones under various conditions and compared the semen quality and quantity after that. They had considered drone viability and the energetic state of sperm by assessing the activity of superoxide dismutase (SOD). This enzyme protects cells against damage caused by superoxide anion (Findlay *et al.* 2008). Woyke (1969) suggested a method for rearing drones up to the prepupal stage. Cobey (2007) concluded that sperm cells are stored in the spermatheca. Sperm concentration is a determining indicator for drone fitness, polyandry and sperm competition (Koeniger *et al.* 2005). According to Abdelkader *et al.* (2013), sperm concentration is $2.2–3.4\times 10^6$ /µL semen, whereas Rhodes *et al.* (2010) concluded it as 3.33×10^6 /µL semen. Further, Collins and Pettis (2001) concluded it as 8.66×10^6; Schlüs *et al.* 2003, 11.9×10^6. Variation in sperm count can be due to different counting methods.

According to Biekowska *et al.* (2011), the temperature at which drones were kept before semen collection greatly influenced the number of spermatozoa. Locke and Peng (1993) elucidated that sperm viability decreases with drone age. They further observed that sperm viability decreases in 4-6 week old drones compared to 2-week-old drones. Additionally, diluent characteristics like pH, osmolarity, nutrients, and handling method influence sperm cell survival. Sperm cell motion depends on the cells' physical environment, temperature, chemical environment, metabolic capacity, and structural integrity. Superoxide dismutase is an antioxidant enzyme involved in the survival of drone spermatozoa (Wegener et al., 2012). Maintaining a balance between reactive oxygen species and degeneration requires an adequate level of SOD (Gavella *et al.* 1996). According to Kurpisz *et al.* (1996), an appropriate level of SOD activity must maintain sperm movement. Baer *et al.* (2009) detected about fifty-seven different types of protein in the seminal fluid of honey bees. Some proteins promote sperm viability, while others protect against microbial attack or reduce oxidative stress in the sperm (Abdelkader et al., 2013).

Chuttong et al., 2019 investigated the production of drones by workers in *Apis mellifera*. Drone production and maintenance require significant investment by honey bee colonies (Page, 1981; Page & Metcalf, 1984; Seeley, 1985; Seeley, 2002; Chuttong et al., 2019). Langstroth had considered them useless consumers. In *A. mellifera* biotypes, drones are produced on a seasonal basis, with commencement in early spring and peaking in late spring to early summer, then declining in mid to late summer (Allen, 1963; Page, 1981; Lee & Winston, 1987). Drone comb production is dependent upon the size of the worker bee population and the time of year (Free and Williams, 1975). According to Seeley and Morse (1976), about 17% of the total comb is given to drone comb. Burgett and Titayavan (2004) reported in *A. florea* that 4.5 per cent of the comb was used for drone production. According to Somana *et al.* (2011), approximately 16.7% of the comb in *A. cerana* is available for drone production in the winter and about 5.7% in the wet summer. Seeley and Morse (1976) reported that about 4.8% of the adult population in a honey bee colony are drones.

CONCLUSION

Mating in honey bees occurs in the drone congregational area, where the queen mates with multiple drones to receive genetic stocks for adding variations. Genetic variations are critical for a species ability to adapt to changing environmental conditions. Variation in the bee population influences productivity, pheromonal secretion, reproductive potential, immunity, and others.

REFERENCES

Adlen, MD I965. The production of queen cups and queen cells in relation to the general development of honeybee colonies and its connection with swarming and supersedure. *J Apic Res,* 4, 121-41.

Alberto Galindo-Cardonaa*, A. Carolina Monmany, Rafiné Moreno-Jackson, Carlos Rivera-Rivera, Carlos Huertas-Dones, Laura Caicedo-Quiroga and Tugrul Giray. Landscape analysis of drone congregation areas of the honey bee, *Apis mellifera. J Insect Sci,* 12

Allen, MD (1963) Drone production in honeybee colonies (*Apis mellifera* L.). *Nature,* 199, 789-90. [http://dx.doi.org/10.1038/199789a0]

Amiri, E, Strand, M, Rueppell, O & Tarpy, D (2017) Queen quality and the impact of honey bee diseases on queen health: potential for interactions between two major threats to colony health. *Insects,* 8, 48. [http://dx.doi.org/10.3390/insects8020048] [PMID: 28481294]

Andersen, D (2004) Improving queen bee production. A report for the Rural Industries Research and Development Corporation. Publication CSE-85A, Rural Industries Research and Development Corporation, Barton, Australia. Available from: https:// rirdc.infoservices.com.au/downloads/04-153 accessed 14 February 2015.

Ben Abdelkader, F, Kairo, G, Tchamitchian, S, Cousin, M, Senechal, J, Crauser, D, Vermandere, JP, Alaux, C, Le Conte, Y, Belzunces, LP, Barbouche, N & Brunet, JL (2014) Semen quality of honey bee drones maintained from emergence to sexual maturity under laboratory, semi-field and field conditions. *Apidologie (Celle),* 45, 215-23.

[http://dx.doi.org/10.1007/s13592-013-0240-7]

Baer, B, Heazlewood, JL, Taylor, NL, Eubel, H & Millar, AH (2009) The seminal fluid proteome of the honeybee *Apis mellifera. Proteomics,* 9, 2085-97.
[http://dx.doi.org/10.1002/pmic.200800708] [PMID: 19322787]

Baudry, E, Solignac, M, Garnery, L, Gries, M, Cornuet, J & Koeniger, N (1998) Relatedness among honeybees (*Apis mellifera*) of a drone congregation. *Proc Biol Sci,* 265, 2009-14.
[http://dx.doi.org/10.1098/rspb.1998.0533]

Berg, S (1991) Investigation on rates of large and small drones at a drone congregation area. *Apidologie (Celle),* 22, 437-8.

Berg, S, Koeniger, N, Koeniger, G & Fuchs, S (1997) Body size and reproductive success of drones (Apis mellifera L). *Apidologie (Celle),* 28, 449-60.
[http://dx.doi.org/10.1051/apido:19970611]

Bienefeld, K (2016) Breeding Success or Genetic Diversity in Honey Bees? *Bee World,* 93, 40-4.
[http://dx.doi.org/10.1080/0005772X.2016.1227547]

Bieńkowska, M, Panasiuk, B, Węgryznowicz, P & Gerula, D (2011) The effect of different thermal conditions on drone semen quality and the number of spermatozoa entering the spermatheca of the queen bee. *J Apic Sci,* 55, 161-8.

Bishop, GH (1920) Fertilization in the honey-bee. I. The male sexual organs: Their histological structure and physiological functioning. *J Exp Zool,* 31, 224-65.
[http://dx.doi.org/10.1002/jez.1400310203]

Brandstaetter, AS, Bastin, F & Sandoz, JC (2014) Honeybee drones are attracted by groups of consexuals in a walking simulator. *J Exp Biol,* 217, jeb.094292.
[http://dx.doi.org/10.1242/jeb.094292] [PMID: 24436379]

Brutscher, LM, Baer, B & Niño, EL (2019) Putative drone copulation factors regulating honey bee (*Apis mellifera*) queen reproduction and health: A review. *Insects,* 10, 8.
[http://dx.doi.org/10.3390/insects10010008] [PMID: 30626022]

Burgett, DM & Titayavan, M (2004) Apis florea F. – colony biometrics in northern Thailand. *Proceeding 8th IBRA Conference on Tropical Beekeeping & VI Encontro sobre Abelhas.* Ribeirao Preto, Brazil. pp. 46-54.

Camazine, S, Cakmak, I, Cramp, K, Fisher, J, Frazier, M & Rozo, A (1998) How healthy are commercially-produced US honey bee queens? *Am Bee J,* 138, 677-80.

Châline, N, Ratnieks, FLW, Raine, NE, Badcock, NS & Burke, T (2004) Non-lethal sampling of honey bee, *Apis mellifera*, DNA using wing tips. *Apidologie (Celle),* 35, 311-8.
[http://dx.doi.org/10.1051/apido:2004015]

Chuttong, B, Buawangpong, N & Burgett, M (2019) Drone Production by the Giant Honey Bee Apis dorsata F. (*Hymenoptera: Apidae*). *Sociobiology,* 66, 475-9.
[http://dx.doi.org/10.13102/sociobiology.v66i3.4355]

Cobey, SW (2007) Comparison studies of instrumentally inseminated and naturally mated honey bee queens and factors affecting their performance. *Apidologie (Celle),* 38, 390-410.
[http://dx.doi.org/10.1051/apido:2007029]

Collet, T, Cristino, AS, Quiroga, CFP, Soares, AEE & Del Lama, MA (2009) Genetic structure of drone congregation areas of Africanized honeybees in southern Brazil. *Genet Mol Biol,* 32, 857-63.
[http://dx.doi.org/10.1590/S1415-47572009005000083] [PMID: 21637465]

Collins, AM & Pettis, JS (2001) Effect of Varroa infestation on semen quality. *Am Bee J,* 141, 590.

Collison, CH (2004) *Basics of Beekeeping.* The University of Pennsylvania, University Park.

Couvillon, MJ, Hughes, WOH, Perez-Sato, JA, Martin, SJ, Roy, GGF & Ratnieks, FLW (2010) Sexual selection in honey bees: colony variation and the importance of size in male mating success. *Behav Ecol,* 21,

520-5.
[http://dx.doi.org/10.1093/beheco/arq016]

Crozier, RH & Page, RE (1985) On being the right size: male contributions and multiple mating in social Hymenoptera. *Behav Ecol Sociobiol,* 18, 105-15.
[http://dx.doi.org/10.1007/BF00299039]

Czekońska, K, Chuda-Mickiewicz, B & Chorbiski, P (2013) The influence of age of honey bee (*Apis mellifera*) drones on the volume of semen and viability of spermatozoa. *J Apic Sci,* 57, 61-6. a

Degrandi-Hoffman, G, Spivak, M & Martin, JH (1993) Role of thermoregulation by nestmates on the development time of honey bee (Hymenoptera: Apidae) queens. *Ann Entomol Soc Am,* 86, 165-72.
[http://dx.doi.org/10.1093/aesa/86.2.165]

Degrandi-Hoffman, G, Watkins, JC, Collins, AM, Loper, GM, Martin, JH, Arias, MC & Sheppard, WS (1998) Queen developmental time as a factor in the Africanization of European honey bee (Hymenoptera: Apidae) populations. *Ann Entomol Soc Am,* 91, 52-8.
[http://dx.doi.org/10.1093/aesa/91.1.52]

Di Pasquale, G, Salignon, M, Le Conte, Y, Belzunces, LP, Decourtye, A & Kretzschmar, A (2013) Influence of pollen nutrition on honey bee health: Do pollen quality and diversity matter? *Public Library of Science One,* 8, 1-13.
[http://dx.doi.org/10.1371/journal.pone.0072016]

Duay, P, De Jong, D & Engels, W (2002) Decreased flight performance and sperm production in drones of the honey bee (*Apis mellifera*) slightly infested by Varroa destructor mites during pupal development. *Genet Mol Res,* 1, 227-32.
[PMID: 14963829]

Estoup, A, Garnery, L, Solignac, M & Cornuet, JM (1995) Microsatellite variation in honey bee (*Apis mellifera* L.) populations: hierarchical genetic structure and test of the infinite allele and stepwise mutation models. *Genetics,* 140, 679-95.
[http://dx.doi.org/10.1093/genetics/140.2.679] [PMID: 7498746]

Fahrenholz, L, Lamprecht, I & Schricker, B (1992) Calorimetric investigations of the different castes of honey bees, Apis mellifera carnica. *J Comp Physiol B,* 162, 119-30.
[http://dx.doi.org/10.1007/BF00398337]

Free, JB & Williams, IH (1975) Factors determining the rearing and rejection of drones by the honeybee colony. *Anim Behav,* 23, 650-75.
[http://dx.doi.org/10.1016/0003-3472(75)90143-8]

Free, JB (1987) *Pheromones of Social Bees.*Comstock Press, Ithaca.

Galindo-Cardona, A 2010. Male Behavior and hybridization of Africanized and European bees, PhD. Dissertation. University of Puerto Rico, Rio Piedras.

Galindo-Cardona, A, Monmany, AC, Moreno-Jackson, R, Rivera-Rivera, C, Huertas-Dones, C, Caicedo-Quiroga, L & Giray, T (2012) Landscape analysis of drone congregation areas of the honey bee, *Apis mellifera. J Insect Sci,* 12, 1-15.
[http://dx.doi.org/10.1673/031.012.12201] [PMID: 23451901]

Galindo-Cardona, A, Monmany, AC, Moreno-Jackson, R, Rivera-Rivera, C, Huertas-Dones, C, Caicedo-Quiroga, L & Giray, T (2012) Landscape analysis of drone congregation areas of the honey bee, *Apis mellifera. J Insect Sci,* 12, 1-15.
[http://dx.doi.org/10.1673/031.012.12201] [PMID: 23451901]

Galindo-Cardona, A, Carolina Monmany, A, Moreno-Jackson, R, Rivera-Rivera, C, Huertas-Dones, C, Caicedo-Quiroga, L & Giray, T (2012) Landscape analysis of drone congregation areas of the honey bee, Apis mellifera. *Journal of Insect Science,* 12, 122.

Galindo-Cardona, A, Monmany, AC, Diaz, G & Giray, T (2015) A landscape analysis to understand the orientation of honey bee (Hymenoptera: Apidae) drones in Puerto Rico. *Environ Entomol,* 44, 1139.1-48.

[http://dx.doi.org/10.1093/ee/nvv099] [PMID: 26314058]

Gary, NE (1962) Chemical mating attractants in the queen honey bee. Science, N. Y., 136, 773-774. Gary, N. E. 1963. Observations of mating behaviour in the honey bee. *J Apic Res,* 2, 3-13.
[http://dx.doi.org/10.1080/00218839.1963.11100050]

Gary, NE (1963) Observations of mating behaviour in the honey bee. *J Apic Res,* 2, 3-13.
[http://dx.doi.org/10.1080/00218839.1963.11100050]

Gençer, HV & Kahya, Y (2011) Are sperm traits of drones (*Apis mellifera* L.) from laying worker colonies noteworthy? *J Apic Res,* 50, 130-7.
[http://dx.doi.org/10.3896/IBRA.1.50.2.04]

Gençer, HV & Firatli, Ç (2005) Reproductive and morphological comparisons of drones reared in queenright and laying worker colonies. *J Apic Res,* 44, 163-7.
[http://dx.doi.org/10.1080/00218839.2005.11101172]

Gerig, L (1971) Wie Drohnen auf K6niginnenattrappen reagieren. Schweiz. *Bienenz,* 94, 558-62.

Petersen, GEL, Fennessy, PF, Van Stijn, TC, Clarke, SM, Dodds, KG & Dearden, PK (2020) Genotyping-b--sequencing of pooled drone DNA for the management of living honeybee (Apis mellifera) queens in commercial beekeeping operations in New Zealand. *Apidologie (Celle),* 51, 545-56.
[http://dx.doi.org/10.1007/s13592-020-00741-w]

Goins, A & Schneider, SS (2013) Drone "quality" and caste interactions in the honey bee, *Apis mellifera* L. *Insectes Soc,* 60, 453-61.
[http://dx.doi.org/10.1007/s00040-013-0310-x]

Gregory, PG & Rinderer, TE (2004) Non-destructive sources of DNA used to genotype honey bee (*Apis mellifera*) queens. *Entomol Exp Appl,* 111, 173-7.
[http://dx.doi.org/10.1111/j.0013-8703.2004.00164.x]

Gries, M & Koeniger, N (1996) Straight forward to the queen: pursuing honeybee drones (Apis mellifera L.) adjust their body axis to the direction of the queen. *J Comp Physiol A Neuroethol Sens Neural Behav Physiol,* 179, 539-44.
[http://dx.doi.org/10.1007/BF00192319]

Harbo, JR & Williams, JL (1987) Effect of above-freezing temperatures on temporary storage of honeybee spermatozoa. *J Apic Res,* 26, 53-5.
[http://dx.doi.org/10.1080/00218839.1987.11100735]

Henryon, M, Liu, H, Berg, P, Su, G, Nielsen, HM, Gebregiwergis, GT & Sørensen, AC (2019) Pedigree relationships to control inbreeding in optimum-contribution selection realise more genetic gain than genomic relationships. *Genet Sel Evol,* 51, 39.
[http://dx.doi.org/10.1186/s12711-019-0475-5] [PMID: 31286868]

Hoage, TR & Kessel, RG (1968) An electron microscope study of the differentiation process during spermatogenesis in the drone honey bee (*Apis mellifera* L.) with particular reference to centriole replication and elimination. *J Ultrastruct Res,* 24, 6-32.
[http://dx.doi.org/10.1016/S0022-5320(68)80014-0] [PMID: 5683704]

Holldobler, B & Bartz, SH (1985) Sociobiology of reproduction in ants. *Prog Zool,* 31, 237-57.

Abou-Shaara, HF & Kelany, MM (2021) A methodology to assist in locating drone congregation area using remote sensing technique. *J Apic Res*
[http://dx.doi.org/10.1080/00218839.2021.1898786]

Abou-Shaara, HF & Kelany, MM (2021) A methodology to assist in locating drone congregation area using remote sensing technique. *J Apic Res*
[http://dx.doi.org/10.1080/00218839.2021.1898786]

Hrassnigg, N & Crailsheim, K (2005) Differences in drone and worker physiology in honeybees (*Apis mellifera*). *Apidologie (Celle),* 36, 255-77.

[http://dx.doi.org/10.1051/apido:2005015]

Jaffé, R & Moritz, RFA (2010) Mating flights select for symmetry in honeybee drones (Apis mellifera). *Naturwissenschaften,* 97, 337-43.
[http://dx.doi.org/10.1007/s00114-009-0638-2] [PMID: 20012931]

Jean-Prost, P (1958) Resumee des obversation sur le vol nuptial des reines abeille. *Proc Int Beekeep Congr Rome,* 404-8.

Johnson, RM, Dahlgren, L, Siegfried, BD & Ellis, MD (2013) Effect of in-hive miticides on drone honey bee survival and sperm viability. *J Apic Res,* 52, 88-95.
[http://dx.doi.org/10.3896/IBRA.1.52.2.18]

Juliana, RANGEL Factors affecting the reproductive health of honey bee (Apis mellifera) drones—a review. *Apidologie.*

Juliana, RANGEL & Adrian, FISHER Factors affecting the reproductive health of honey bee (Apis mellifera) drones—a review. *Apidologi.*

Kerr, WE & Bueno, D (1990) 1970. Natural crossing between *Apis mellifera* adansonii and *Apis mellifera* ligustica. Evolution, 24, 145-148. The role of the mating sign in honey bees, *Apis mellifera* L.: does it hinder or promote multiple mating? *Anim Behav,* 39, 444-9.

King, M, Eubel, H, Millar, AH & Baer, B (2011) Proteins within the seminal fluid are crucial to keep sperm viable in the honeybee Apis mellifera. *J Insect Physiol,* 57, 409-14.
[http://dx.doi.org/10.1016/j.jinsphys.2010.12.011] [PMID: 21192944]

Koeniger, G, Koeniger, N, Ellis, J & Connor, L (2015) *Mating biology of honey bees (Apis mellifera).* Wicwas Press, Kalamazoo.

Koeniger, N, Koeniger, G & Pechhacker, H (2005) The nearer the better? Drones (*Apis mellifera*) prefer nearer drone congregation areas. *Insectes Soc,* 52, 31-5.
[http://dx.doi.org/10.1007/s00040-004-0763-z]

Koeniger, G (1990) The role of the mating sign in honey bees, *Apis mellifera* L.: does it hinder or promote multiple mating? *Anim Behav,* 39, 444-9.
[http://dx.doi.org/10.1016/S0003-3472(05)80407-5]

Koeniger, G (1986) Mating sign and multiple mating in the honeybee. *Bee World,* 67, 141-50.
[http://dx.doi.org/10.1080/0005772X.1986.11098892]

Koeniger, G (1988) Mating behavior of honey bees.*Africanized Honey Bees and Bee Mites* Halstead Press, New York 167-72.

Koeniger, G (1990) The role of the mating sign in honey bees, *Apis mellifera* L.: does it hinder or promote multiple mating? *Anim Behav,* 39, 444-9.
[http://dx.doi.org/10.1016/S0003-3472(05)80407-5]

Koeniger, N, Koeniger, G & Pechhacker, H (2005) The nearer the better? Drones (*Apis mellifera*) prefer nearer drone congregation areas. *Insectes Soc,* 52, 31-5.
[http://dx.doi.org/10.1007/s00040-004-0763-z]

Koeniger, N, Koeniger, G, Gries, M & Tingek, S (2005) Drone competition at drone congregation areas in four *Apis* species. *Apidologie (Celle),* 36, 211-21. a
[http://dx.doi.org/10.1051/apido:2005011]

Koeniger, N, Koeniger, G & Pechhacker, H (2005) The nearer the better? Drones (*Apis mellifera*) prefer nearer drone congregation areas. *Insectes Soc,* 52, 31-5. b
[http://dx.doi.org/10.1007/s00040-004-0763-z]

Laidlaw, HH, Jr & Page, RE, Jr (1984) Polyandry in honey bees (*Apis mellifera* L.) sperm utilization and intracolony genetic relationships. *Genetics,* 108, 985-97.
[http://dx.doi.org/10.1093/genetics/108.4.985] [PMID: 17246245]

Laidlaw, HH (1987) Instrumental insemination of honeybee queens: its origin and development. *Bee World,* 68, 17-36.

Laidlaw, HH, Jr & Page, RE, Jr (1984) Polyandry in honey bees (*Apis mellifera* L.): sperm utilization and intracolony genetic relationships. *Genetics,* 108, 985-97.
[http://dx.doi.org/10.1093/genetics/108.4.985] [PMID: 17246245]

Lee, PC & Winston, ML (1987) Effects of reproductive timing and colony size on the survival, offspring colony size and drone production in the honey bee (Apis mellifera). *Ecol Entomol,* 12, 187-95.
[http://dx.doi.org/10.1111/j.1365-2311.1987.tb00997.x]

Lensky, Y & Demter, M (1985) Mating flights of the queen honeybee (*Apis mellifera*) in a subtropical climate. *Comp Biochem Physiol A Comp Physiol,* 81, 229-41.
[http://dx.doi.org/10.1016/0300-9629(85)90127-6]

Locke, SJ & Peng, YS (1993) The effects of drone age, semen storage and contamination on semen quality in the honey bee (Apis mellifera). *Physiol Entomol,* 18, 144-8.
[http://dx.doi.org/10.1111/j.1365-3032.1993.tb00461.x]

Loper, GM, Wolf, WW & Taylor, OR, Jr (1992) Honey bee drone flyways and congregationareas radar observations. *J Kans Entomol Soc,* 65, 223-30.

Loper, GM, Wolf, WW & Taylor, OR, Jr (1987) Detection and monitoring of honeybee drone congregation areas by radar. *Apidologie (Celle),* 18, 163-72.
[http://dx.doi.org/10.1051/apido:19870206]

Loper, GM, Wolf, WW & Taylor, OR (1992) Honey-bee drone flyways and congregation areas – radar observations. *J Kans Entomol Soc,* 65, 223-30.

Mackensen, O (1955) Experiments in the technique of artificial insemination of queen bees. *J Econ Entomol,* 48, 418-21.
[http://dx.doi.org/10.1093/jee/48.4.418]

Mackensen, O & Roberts, WC (1948) A manual for the artificial insemination of queen bees. USDA. *Bureau of Entomology and Plant Quarantine,* ET-250, 1-33.

Plate, M, Bernstein, R & Hoppe, A (2019) The importance of controlled mating in honeybee breeding. *Genet Sel Evol,* 51, 74.
[http://dx.doi.org/10.1186/s12711-019-0518-y]

Metz, B & Tarpy, D (2019) Reproductive senescence in drones of the honey bee (*Apis mellifera*). *Insects,* 10, 11.
[http://dx.doi.org/10.3390/insects10010011] [PMID: 30626026]

Meuwissen, TH (1997) Maximizing the response of selection with a predefined rate of inbreeding. *J Anim Sci,* 75, 934-40.
[http://dx.doi.org/10.2527/1997.754934x] [PMID: 9110204]

Moritz, RFA, Kryger, P & Allsopp, MH (1996) Competition for royalty in bees. *Nature,* 384, 31-1.
[http://dx.doi.org/10.1038/384031a0]

Moritz, RFA (1989) *The instrumental insemination of the queen bee* Apimondia Publishing House, Bucharest 22-8.

Neumann, P & Moritz, RFA (2000) Testing genetic variance hypotheses for the evolution of polyandry in the honeybee (Apis mellifera L.). *Insectes Soc,* 47, 271-9.
[http://dx.doi.org/10.1007/PL00001714]

Nguyen, VN (1995)

Nolan, WJ (1932) *Breeding the honeybee under controlled conditions.*US Department of Agriculture, New York.

Nur, Z, Seven-Cakmak, S, Ustuner, B, Cakmak, I, Erturk, M, Abramson, CI, Sağirkaya, H & Soylu, MK

(2012) The use of the hypo-osmotic swelling test, water test, and supravital staining in the evaluation of drone sperm. *Apidologie (Celle),* 43, 31-8.
[http://dx.doi.org/10.1007/s13592-011-0073-1]

Oertel, E (1956) Observations on the flight of the drone hone bees. *Ann Entomol Soc Am,* 49, 497-500.
[http://dx.doi.org/10.1093/aesa/49.5.497]

Page, RE, Jr (1981) Protandrous reproduction in honey bees. *Environ Entomol,* 10, 359-62.
[http://dx.doi.org/10.1093/ee/10.3.359]

Page, RE, Jr & Metcalf, RA (1984) A population investment sex ratio for the honey bee (*Apis mellifera* L.). *Am Nat,* 124, 680-702.
[http://dx.doi.org/10.1086/284306]

Page, RE, Jr & Metcalf, RA (1982) Multiple mating, sperm utilization, and social evolution. *Am Nat,* 119, 263-281, 263-281.
[http://dx.doi.org/10.1086/283907]

Page, RE, Jr (1980) The evolution of multiple mating behavior by honey bee queens (*Apis mellifera* L.). *Genetics,* 96, 263-73.
[http://dx.doi.org/10.1093/genetics/96.1.263] [PMID: 7203010]

Page, RE, Jr (1981) Protandrous Reproduction in Honey Bees. *Environ Entomol,* 10, 359-62.
[http://dx.doi.org/10.1093/ee/10.3.359]

Page, RE, Jr (1986) Sperm utilization in social insects. *Annu Rev Entomol,* 31, 297-320.
[http://dx.doi.org/10.1146/annurev.en.31.010186.001501]

Page, RE, Jr & Peng, CYS (2001) Aging and development in social insects with emphasis on the honey bee, *Apis mellifera* L. *Exp Gerontol,* 36, 695-711.
[http://dx.doi.org/10.1016/S0531-5565(00)00236-9] [PMID: 11295509]

Pain, J & Ruttner, F (1963) Les extrait des glandes mandibulaires des reines d'abeilles attirent les males, lors du vol nuptial. *CR Acad Sci,* 256, 512-5.

Palmer, KA & Oldroyd, BP (2000) Evolution of multiple mating in the genus *Apis. Apidologie (Celle),* 31, 235-48.
[http://dx.doi.org/10.1051/apido:2000119]

Peng, Y, Grassl, J, Millar, AH & Baer, B (2016) Seminal fluid of honeybees contains multiple mechanisms to combat infections of the sexually transmitted pathogen Nosema apis *Proc R Soc Lond Biol Sci* 283 (1823), 20151785.
[http://dx.doi.org/10.1098/rspb.2015.1785]

Rangel, J, Keller, JJ & Tarpy, DR (2013) The effects of honey bee (*Apis mellifera* L.) queen reproductive potential on colony growth. *Insectes Soc,* 60, 65-73.
[http://dx.doi.org/10.1007/s00040-012-0267-1]

Rhodes, JW (2002) The drone hone bees: Rearing and maintenance.

Rhodes, JW (2008) Semen production in drone honeybees [online]. Publication No. 08/130. Rural Industries Research and Development Corporation, Barton, Australia. Available from: https://rirdc.infoservices.com (Accessed 5 June 2018).

Rhodes, JW, Harden, S, Spooner-Hart, R, Anderson, DL & Wheen, G (2011) Effects of age, season and genetics on semen and sperm production in *Apis mellifera* drones. *Apidologie (Celle),* 42, 29-38.
[http://dx.doi.org/10.1051/apido/2010026]

Roberts, WC (1944) Multiple mating of queen bees proved by progeny and flight tests. *Glean Bee Cult,* 72, 255-60.

Roberts, WC & Mackensen, O (1951) Breeding improved honey bees. *Am Bee J,* 91, 473-5.

Robinson, GE, Fernald, RD & Clayton, DF (2008) Genes and social behavior. *Science,* 322, 896-900.

[http://dx.doi.org/10.1126/science.1159277] [PMID: 18988841]

Robinson, GE & Page, RE, Jr (1988) Genetic determination of guarding and undertaking in honey-bee colonies. *Nature,* 333, 356-8.
[http://dx.doi.org/10.1038/333356a0]

Rousseau, A, Fournier, V & Giovenazzo, P (2015) *Apis mellifera* (Hymenoptera: Apidae) drone sperm quality in relation to age, genetic line, and time of breeding. *Can Entomol,* 147, 702-11.
[http://dx.doi.org/10.4039/tce.2015.12]

Rousseau, A & Giovenazzo, P (2016) Optimizing drone fertility with spring nutritional supplements to honey bee (Hymenoptera: Apidae) colonies. *J Econ Entomol,* 109, 1009-14.
[http://dx.doi.org/10.1093/jee/tow056] [PMID: 27018435]

Rousseau, A & Giovenazzo, P Université Laval, 2480 boulevard Hochelaga, Ville de Québec, Québec, Canada G1V 0A6; and Centre de recherche en sciences animales de Deschambault, 120-A, chemin du Roy, \ Deschambault, Québec, Canada G0A 1S0.

Rowland, CM & McLellan, AR (1987) Seasonal changes of drone numbers in a colony of the honeybee, *Apis mellifera. Ecol Modell,* 37, 155-66.
[http://dx.doi.org/10.1016/0304-3800(87)90023-8]

Ruttner, H (1976) Investigations on the flight activity and the mating Behavior of drones. VI. Flight on and over mountain ridges. *Apidologie (Celle),* 7, 331-41.
[http://dx.doi.org/10.1051/apido:19760404]

Ruttner, F & Ruttner, H (1963) Die Flugaktivitfit und das Paarungsverhalten der Drohnen. 297-301.

Ruttner, F & Ruttner, H (1965) Untersuchungen fiber die Flugaktivitfit und das Paarungsverhalten der Drohnen. *Z Bienenforsch,* 8, 1-9.

Ruttner, F (1966) The life and flight activity of drones. *Bee World,* 47, 93-100.
[http://dx.doi.org/10.1080/0005772X.1966.11097111]

Ruttner, H (1976) Investigations on the flight activity and the mating Behavior of drones. VI. Flight on and over mountain ridges. *Apidologie (Celle),* 7, 331-41.
[http://dx.doi.org/10.1051/apido:19760404]

Ruttner, H & Ruttner, F (1972) Untersuchungen €uber die Flugaktivit€at und das Paarungsverhalten der Drohnen. V.- Drohnensammelpl€atze und Paarungsdistanz. *Apidologie (Celle),* 3, 203-32.
[http://dx.doi.org/10.1051/apido:19720301]

Scheiner, R, Abramson, CI, Brodschneider, R, Crailsheim, K, Farina, WM, Fuchs, S, Grünewald, B, Hahshold, S, Karrer, M, Koeniger, G, Koeniger, N, Menzel, R, Mujagic, S, Radspieler, G, Schmickl, T, Schneider, C, Siegel, AJ, Szopek, M & Thenius, R (2013) Standard methods for behavioural studies of *Apis mellifera. J Apic Res,* 52, 1-58.
[http://dx.doi.org/10.3896/IBRA.1.52.4.04]

Schlüns, H, Moritz, RFA, Neumann, P, Kryger, P & Koeniger, G (2005) Multiple nuptial flights, sperm transfer and the evolution of extreme polyandry in honeybee queens. *Anim Behav,* 70, 125-31.
[http://dx.doi.org/10.1016/j.anbehav.2004.11.005]

Schlüns, H, Schlüns, EA, van Praagh, J & Moritz, RFA (2003) Sperm numbers in drone honeybees (*Apis mellifera*) depend on body size. *Apidologie (Celle),* 34, 577-84.
[http://dx.doi.org/10.1051/apido:2003051]

Schlüns, H, Schlüns, EA, van Praagh, J & Moritz, RFA (2003) Sperm numbers in drone honeybees (*Apis mellifera*) depend on body size. *Apidologie (Celle),* 34, 577-84.
[http://dx.doi.org/10.1051/apido:2003051]

Seeley, TD (1985) *Honeybee Ecology A Study of Adaptation in Social Life.*Princeton University Press.
[http://dx.doi.org/10.1515/9781400857876]

Seeley, TD (1985) *Honeybee Ecology.*Princeton University Press.

[http://dx.doi.org/10.1515/9781400857876]

Seeley, TD (2002) The effect of drone comb on a honey bee colony's production of honey. *Apidologie (Celle),* 33, 75-86.
[http://dx.doi.org/10.1051/apido:2001008]

Seeley, TD & Morse, RA (1976) The nest of the honey bee (Apis mellifera L.). *Insectes Soc,* 23, 495-512.
[http://dx.doi.org/10.1007/BF02223477]

Seeley, TD (1985) *Honeybee Ecology A Study of Adaptations in Social Life.*Princeton University Press, Princeton, New Jersey.
[http://dx.doi.org/10.1515/9781400857876]

Sherman, PW, Seeley, TD & Reeve, HK (1988) Parasites, pathogens, and polyandry in social Hymenoptera. *Am Nat,* 131, 602-10.
[http://dx.doi.org/10.1086/284809]

Slone, JD, Stout, TL, Huang, ZY & Schneider, SS (2012) The influence of drone physical condition on the likelihood of receiving vibration signals from worker honey bees, *Apis mellifera. Insectes Soc,* 59, 101-7.
[http://dx.doi.org/10.1007/s00040-011-0195-5]

Smith, ML, Ostwald, MM, Loftus, JC & Seeley, TD (2014) A critical number of workers in a honeybee colony triggers investment in reproduction. *Naturwissenschaften,* 101, 783-90.
[http://dx.doi.org/10.1007/s00114-014-1215-x] [PMID: 25142633]

Snodgrass, RE (1956) *Anatomy of the Honey Bee* Cornell University Press, Ithaca 290-8.

Somana, W, Chanbang, Y, Kulsarin, J & Burgett, M (2011) Biometrics of the natural nest of the eastern honey bee (Apis cerana F.) as observed in northern Thailand. *Chiang Mai University Journal of Agriculture,* 27, 219-28. [In Thai].

Strang, GE (1970) A study of honey bee drone attraction in mating response. *J Econ Entomol,* 63, 641-5.
[http://dx.doi.org/10.1093/jee/63.2.641]

Stürup, M, Baer-Imhoof, B, Nash, DR, Boomsma, JJ & Baer, B (2013) When every sperm counts: factors affecting male fertility in the honeybee *Apis mellifera. Behav Ecol,* 24, 1192-8.
[http://dx.doi.org/10.1093/beheco/art049]

Stürup, M, Baer-Imhoof, B, Nash, DR, Boomsma, JJ & Baer, B (2013) When every sperm counts: factors affecting male fertility in the honeybee *Apis mellifera. Behav Ecol,* 24, 1192-8.
[http://dx.doi.org/10.1093/beheco/art049]

Stürup, M, Baer-Imhoof, B, Nash, DR, Boomsma, JJ & Baer, B (2013) When every sperm counts: factors affecting male fertility in the honeybee *Apis mellifera. Behav Ecol,* 24, 1192-8.
[http://dx.doi.org/10.1093/beheco/art049]

Taha, ELKA & Alqarni, AS (2013) Morphometric and reproductive organs characters of *Apis mellifera* jemenitica drones in comparison to *Apis mellifera* carnica. *Int J Sci Eng Res,* 4, 411-5.

Tarpy, DR & Nielsen, DI (2002) Sampling error, efective paternity, and estimating the genetic structure of honey bee colonies (Hymenoptera: Apidae). *Ann Entomol Soc Am,* 95, 513-28.
[http://dx.doi.org/10.1603/0013-8746(2002)095[0513:SEEPAE]2.0.CO;2]

Tarpy, DR & Page, RE, Jr (2000) No behavioral control over mating frequency in queen honey bees (*Apis mellifera* L.): implications for the evolution of extreme polyandry. *Am Nat,* 155, 820-7.
[http://dx.doi.org/10.1086/303358] [PMID: 10805647]

Tarpy, DR, Keller, JJ, Caren, JR & Delaney, DA (2012) Assessing the mating 'health' of commercial honey bee queens. *J Econ Entomol,* 105, 20-5.
[http://dx.doi.org/10.1603/EC11276] [PMID: 22420250]

Tarpy, DR & Page, RE, Jr (2000) No behavioral control over mating frequency in queen honey bees (*Apis mellifera* L.): implications for the evolution of extreme polyandry. *Am Nat,* 155, 820-7.

[http://dx.doi.org/10.1086/303358] [PMID: 10805647]

Thornhill, R & Alcock, J (1983) *The Evolution of Insect Mating Systems.*Harvard University Press, Cambridge, Massachusetts.
[http://dx.doi.org/10.4159/harvard.9780674433960]

Tofilski, A & Kopel, J (1996) The influence of Nosema apis on maturation and flight activity of honey bee drones. *Pszcz Zesz Nauk,* 40, 55-60.

vanEngelsdorp, D & Meixner, MD (2010) A historical review of managed honey bee populations in Europe and the United States and the factors that may affect them. *J Invertebr Pathol,* 103 (Suppl. 1), S80-95.
[http://dx.doi.org/10.1016/j.jip.2009.06.011] [PMID: 19909973]

Verbeek, B (1976) Investigation of the flight activity of young honeybee queens under continental and insular conditions by means of photoelectronic control. *Apidologie (Celle),* 7, 151-68.
[http://dx.doi.org/10.1051/apido:19760205]

Villar, G, Wolfson, MD, Hefetz, A & Grozinger, CM (2018) Evaluating the Role of Drone-Produced Chemical Signals in Mediating Social Interactions in Honey Bees (Apis mellifera). *J Chem Ecol,* 44, 1-8.
[http://dx.doi.org/10.1007/s10886-017-0912-2] [PMID: 29209933]

Weiss, K (1962) Untersuchungen fiber die Drohnenerzeugung im Bienenvolk. *Archf Bienenk,* 39, 1-7.

Wilson, EO (1971) *The Insect Societies.*Harvard University Press, Cambridge.

Winston, ML (1987) *The biology of the honey bee.*Harvard University Press.

Winston, ML & Taylor, OR (1980) Factors preceding queen rearing in the Africanized honeybee (*Apis mellifera*) in South America. *Insectes Soc,* 27, 289-304.
[http://dx.doi.org/10.1007/BF02223722]

Winston, ML (1980) Swarming, afterswarming, and reproductive rate of unmanaged honeybee colonies (Apis mellifera). *Insectes Soc,* 27, 391-8.
[http://dx.doi.org/10.1007/BF02223731]

Winston, ML (1987) *The Biology of the Honey Bee.*Harvard University Press, Cambridge.

Woyke, J & Ruttner, F (1958) An anatomical study of the mating process in honey bee. *Bee World,* 39, 3-18.
[http://dx.doi.org/10.1080/0005772X.1958.11095028]

Woyke, J (1960) Naturalne i sztuczne unasienianie matek pszczelich. *Pszcz Zesz Nauk,* 4, 183-275.

Woyke, J (1962) Natural and artificial insemination of queen bees. *Bee World,* 43, 21-5.
[http://dx.doi.org/10.1080/0005772X.1962.11096922]

Woyke, J (1983) Dynamics of entry of spermatozoa into the spermatheca of instrumentally inseminated queen honey bees. *J Apic Res,* 22, 150-4.
[http://dx.doi.org/10.1080/00218839.1983.11100579]

Woyke, J (2008) Why the eversion of the endophallus of honey bee drone stops at the partly everted stage and significance of this. *Apidologie (Celle),* 39, 627-36.
[http://dx.doi.org/10.1051/apido:2008046]

Woyke, J (1960) Naturalne i sztuczne unasienianiematek pszczelich. *Pszczel Zesz Nauk,* 4, 183-275. [Natural and artificial insemination of queen honeybees].

Woyke, J (1962) Natural and artificial insemination of queen honeybees. *Bee World,* 43, 21-5.
[http://dx.doi.org/10.1080/0005772X.1962.11096922]

Woyke, J & Jasiński, Z (1978) Influence of age of drones on the results of instrumental insemination of honeybee queens. *Apidologie (Celle),* 9, 203-12.
[http://dx.doi.org/10.1051/apido:19780304]

Zaitoun, S, Al-Majeed Al-Ghzawi, A & Kridli, R (2009) Monthly changes in various drone characteristics of *Apis mellifera ligustica* and *Apis mellifera syriaca. Entomol Sci,* 12, 208-14.

[http://dx.doi.org/10.1111/j.1479-8298.2009.00324.x]

Zaitoun, S, Al-Majeed Al-Ghzawi, A & Kridli, R (2009) Monthly changes in various drone characteristics of *Apis mellifera ligustica* and *Apis mellifera syriaca*. *Entomol Sci,* 12, 208-14.
[http://dx.doi.org/10.1111/j.1479-8298.2009.00324.x]

Zmarlicki, C & Morse, RA (1963) Drone congregation areas. *J Apic Res,* 2, 64-6.
[http://dx.doi.org/10.1080/00218839.1963.11100059]

CHAPTER 6

Artificial Methods of Drone Rearing in Apis Mellifera and the Role of Drones in Quality Improvement

Abstract: A few protocols are available for artificial drone rearing under controlled conditions within or outside the honey bee colony. The initial development phase of the drone development can be completed on royal jelly exclusively. Still, on worker jelly, complete drop development can be carried out, as witnessed by a few researchers. Sugars accelerate the drone's growth and support the passage of the drone through certain embryonic and post-embryonic stages. The drones contribute patrilineal genomic content that influences the overall strength of the colony, colonial behaviour, productivity, and other characteristics. The current chapter is attributed to some available artificial haploid/diploid drone-rearing methods and the influence of multiple patrilineal genomic content contributions to other honey bees.

Keywords: Artificial drone rearing, Genetic quality improvement.

INTRODUCTION

Artificial drone rearing is not a common practice in beekeeping as the male honey bees are not directly involved in the colony's productivity. Therefore, very limited explorations are available that are attributed to the development of the drone honey bee, although male honey bees are essential for genetics, embryology, and insect nutrition. Different researchers had tried to rear the drone outside the colony, but minimal information is available about the drones. Takeuchi et al., 1972 reared the diploid drones from the larvae fed on royal jelly mixed with extra sugar, and they succeeded in getting adults at age 2, 4, or 5 days. In total, they could rear about 13 adults artificially. However, the royal jelly could not sustain the complete development of the drone bees.

Rhein (1951) tried to rear the drones on the worker jelly and could get pupae and adults artificially. They concluded that the chemical composition of the drone jelly is similar to that of worker jelly. Jay (1959) attempted to feed the drones royal jelly but did not obtain any drone adults. Woyke (1965) also used royal jelly for diploid drone development. He concluded that drones were challenging to rear compared to queens and further reported that drone larvae pupate if fed on mixed food of older larvae rather than royal jelly exclusively (Woyke, 1963a, 1963b). During the drone larval development, the chemical composition of the drone jelly changes like that of the worker jelly (Gontarski, 1949; Haydak, 1957; Matsuka et al., 1973).

The older drone larvae's diet contains a considerable quantity of sugar. Furthermore, in worker jelly, sugar is known to promote pupation in worker larvae (Shuel and Dixon, 1968). Takeuchi et al., 1972, tried to feed 5-day-old drone honey bee larvae with royal jelly modified with extra sugar to study if sugar could enhance pupation rate and adult emergence. They tried to alter drone jelly by adding 0.8 parts of a 60% sugar solution. They kept the modified diet below 5 degrees Celsius until it was used. For artificial rearing, first, they selected eggs laid in an empty comb for 24 hrs. and kept the comb within the colony for 2–5 days. Then, twenty-five drone larvae were reared by putting them in a plastic vessel containing an experimental diet. The plastic vessel had been placed in a desiccator with controlled humidity. Woyke (l963b) found that the frequency of diet renewal was beneficial to the growth of drone honey bees. At the pre-pupal stage, larvae were transferred to a dish made up of paper for pupation after washing with lukewarm water at 35°C. The relative humidity was maintained at 82% with the help of a supersaturated solution of Na_2SO_4. They followed techniques described by Sasaki and Okada (1972). Further, Takeuchi *et al.* (1972) speculated that drones could be reared in incubators during the latter part of their larval life. Royal jelly can provide the required nutrients for drone development. They further reported that adding sugar increases larvae's body weight and can influence pupation (Shuel and Dixon, 1968). Some components of royal jelly induce inhibition of larval growth. Takeuchi et al., 1972 did not get adults if they tried to feed drone larvae on royal jelly exclusively and, further, they noticed a remarkable reduction in weight in the pre-pupal and pupal phases. Dietz (1966) tried to rear the drones on royal jelly diluted with diet from an older larval stage. According to Jay (1965), young drones were sensitive to the environmental conditions maintained for rearing (Takeuchi et al., 1972).

Woyke, 1962, tested drone-rearing techniques with the best results when diploid drone larvae of 2-3 days old were reared in the incubator. Generally, the diploid drones are eaten after a few hours of hatching by worker honey bees (Woyke, 1962; Woyke, 1963a; Woyke, 1963b). However, the diploid drones are viable and can be reared easily outside the colony in an incubator up to pre-pupal and pupal stages (Woyke, 1963c, 1965b; Woyke, 1965a). The workers eat the diploid drones if laid in worker and drone cells but are more protected in queen cells (Woyke,

1965c; Woyke, 1965d). Nevertheless, no diploid drone cell is reared in queen cells as most of them fall out of the cells.

Woyke, 1962, tried to rear the drone by different methods. The first method includes the artificial rearing of drones for the first two days in an incubator and transferring them to the colony. For that, the worker comb with hatching eggs laid by the queen has been taken out. Afterward, the comb is wrapped in a moist towel or inserted in an isolator with water at the bottom. The rearing had been carried out at a temperature of 34.5°C. Each brood has been examined every 3 hours. After hatching, newly hatched larvae have been grafted into drone cells. The larvae can be placed in an incubator at 34.5°C with a relative humidity of 95–100%. The 2-3 day-old larvae had been transferred to the drone cells. The survival rate was checked two days later, and sex was determined after capping. The sealed brood had been protected with wire gauze.

The second method they have adopted is rearing in colonies throughout the larval life. They had maintained other conditions similar to the first method. Worker bees have been observed preparing convex seals on worker cells containing drone larvae. It has been further noted that bees seal only isolated cells. It has also been speculated that no adult diploid drone emerged from worker cells. The diploid drone larvae can survive for one day or more. They had speculated that a relatively low percentage of larvae reared in incubation could survive in the colony. After being cultivated for 2-3 days in incubators, they concluded that drone cells exhibit a low survival rate.

Their findings indicated that young diploid larvae could be separated and transferred back to a colony where they could be reared like normal diploid females. Woyke, in 1962, assessed the normal survival rate of haploid and diploid drones in different environmental conditions. They compared the haploid drone survival rate during the spring, summer, and autumn seasons. They had considered a haploid drone in a queened and queen-less colony. Woyke (1965) demonstrated that many diploid drones are not eaten by workers when they are located in queen cells.

Polaczek *et al.* (2000) conducted artificial insemination of queens of *Apis mellifera carnica* with the semen of their sons. They also attempted to rear diploid drone offspring using two established techniques, including elaborate laboratory manipulations and small mating nuclei, which facilitate drone raising to adulthood. The ploidy level of drones has been determined by using DNA microsatellites. Furthermore, they reported that elaborative techniques reared all drones, and 90% of drones raised in a small mating nucleus were diploid. The abovementioned method makes it easy to get diploid drones without sophisticated equipment. In honey bees, sex is determined by a single locus. The honey bees, which are heterozygous at the sex locus, develop into females, whereas eggs with hemi or homozygous conditions develop into haploid or diploid males. Insemination is routinely used to get diploid drones like mothers, sons, sisters and brothers. Polaczek *et al.* (2000) suggested a new technique for diploid drone rearing and used DNA microsatellites for reliability testing. Cannibalism in diploid drones is seasonal and dependent on the colony size (Polaczek *et al.* 2000).

GENOMIC CONTRIBUTION OF DRONES TO QUALITY IMPROVEMENT

In the honey bee colony, maternal genomic content inheritance is definite. Still, paternal genomic content inheritance is difficult to determine, even when mating is controlled, as single drone mating negatively affects mating or colony fitness. Petersen et al., 2020, analysed 388 living queens from two beekeeping operations using Genotyping-by-Sequencing (GBS) and a single-nucleotide polymorphism (SNPs) method Tassel 5 and Stacks. In stacks, they could discover more SNPs (29,000). Although beekeeping is an important sector of agriculture, genomic quality improvement is currently lagging and is difficult to control. The reasons are biological features of honey bees, like mating behaviour and limited technologies for getting the desired phenotype and genotype.

Furthermore, for quality improvement, restricted mating should be avoided, and where artificial insemination is performed, it should not be with a single drone, which would restrict the quality improvement and survivability of the colony. Therefore, managing genetic diversity at the population level is key for honey bee breeding programs to maintain a viable population (Bienefeld 2016; Meuwissen 1997). Selection strategies for genetic gain are shown to realise more than other schemes based on genomic information (Henryon et al., 2019).

As the queen mates with multiple diverse drones, complete and correct pedigrees are challenging to obtain in honey bees. Therefore, genomic information is required for the management of genetic diversity. This problem can be resolved by genotyping living queens using wing clippings and larval exuviae (Chaline *et al.* 2004; Gregory and Rinderer 2004). This non-destructive source of DNA requires the implementation of management techniques. According to Petersen *et al.* (2017), a standard protocol for genotyping of queen based on pooled drone samples can be used across an entire industry. Petersen et al., 2020 carried out Genotyping-by-Sequencing (GBS) on a set of samples from two beekeeping operations.

Further, *Apis mellifera* Linnaeus; Hymenoptera: The Apidae queen's life expectancy depends on the number of sperm she receives while mating with drones during the nuptial flight. Rousseau *et al.* (2015) evaluated drone reproductive qualities during the queen-rearing season. Semen volume and sperm viability were assessed in drones for 14, 21, and 35 days. Further, they could observe the drone genetic line, age, or breeding time on sperm viability. European honey bees, *Apis mellifera* Linnaeus (Hymenoptera: Apidae), exhibit a polyandrous mating system (Koeniger 1990; Tarpy and Page 2000). Virgin young queens execute one to three mating flights in drone congregation areas (Tarpy and Page 2000; Schlüns *et al.* 2005).

According to Estoup *et al.* (1995), about 14 drones mate with the queen and then die. After mating, the queen returns with about 80–90 million spermatozoa in her lateral oviducts (Woyke 1962). About 4–7 million received sperm move to her spermatheca, where they remain stored until used by the queen (Laidlaw and Page 1984; Roberts and Mackensen 1951; Woyke 1962).

Some reports describe the queen as having early supersedure, unexplained death, premature drone egg laying, and egg laying interruption (Camazine *et al.* 1998; Rhodes 2008; vanEnglesdorp and Meisner 2010). Tarpy *et al.* (2012) reported that queen quality influences the production of a large amount of viable brood. Further, they concluded that mated queens possess 4.37 million stored sperms in their spermathecae, with an average viability of 83.7%. Rhodes *et al.* (2010) investigated the sperm quality of honeybee drones. They concluded that poor drone quality could also contribute to a low number of sperm in the spermatheca of commercially produced queens. Further, multiple factors influence the mating health of a queen, including rearing conditions or queen age at mating (Cobey 2007). Furthermore, male number and sperm quality influence queen mating success in DCA (Cobey 2007; Nur *et al.* 2012).

Various researchers have speculated that drone age, rearing date, and genetic origin influence semen properties (Woyke and Jasinski 1978; Locke and Peng 1993; Zaitoun *et al.* 2009). According to Woyke and Jasinski (1978), the number of spermatozoa reaching the spermatheca tends to decrease with an increase in the queen's age. They further reported about 4097 million spermatozoa in queens inseminated with the semen of two-week-old drones compared with 3175 million in queens inseminated with the semen of nine-week-old drones. Zaitoun *et al.* (2009) concluded that drones reared in May possess more weight, a higher sperm count, and a higher fertility level than drones raised in the rest of the year. Drone genetics influences sperm production, properties and semen volume (Rhodes *et al.* 2010). Rousseau *et al.* (2015) reported that age and rearing date influence the semen volume, not sperm count or viability. Rhodes *et al.* (2010) concluded that the genetics of drones affect semen production and semen volume in drones. They had identified the drone production volume of semen. Adequate queen fecundation is proportional to the volume of semen produced by each drone and the migration of sperms into her spermatheca in a dose-dependent method (Cobey 2007).

Sexual maturity is reached in drones when sperm migrates from the testes to the seminal vesicles and the mucous glands fully develop (Rhodes 2008). A few explorations indicate that drones mature at 10–21 days, and semen is suitable for queen insemination (Woyke and Jasinski 1978; Harbo and Williams 1987). In field conditions, the estimated life span of A. mellifera drones is 20–40 days (Page and Peng et al. 2001). It has been reported that the mean sperm count was about 1.80±1.65 million (Andersen 2004; Koeniger *et al.* 2005; Rhodes *et al.* 2010; Gençer and Kahya 2011; Nur *et al.* 2012). Honey bee colonies are rarely pollen-starved, but there is a variation in the availability, abundance, type, and diversity of pollen (Di Pasquale et al., 2013). In addition, according to Stürup *et al.* (2013), temperature increases from 35 °C to 39 °C may influence sperm viability.

Due to complex genetic and reproductive characteristics, genetic selection is difficult to apply to honey bees (Apis mellifera). Therefore, the queen is responsible for the overall genetics of each colony.

CONCLUSION

Artificial drone honey bee production is a commendable protocol for improving honey bee colonies. Minimal details are available that provide information of specific processes. Worker jelly and drone jelly provide a suitable medium for the rearing of drone honey bees.

REFERENCES

Andersen, D (2004) Improving queen bee production. A report for the Rural Industries Research and Development Corporation. Publication CSE-85A, Rural Industries Research and Development Corporation, Barton, Australia. Avilable From: https:// rirdc.infoservices.com.au/downloads/04-153

Polaczek, Benedikt, Neumann, Peter, Schricker, Burkhard & Moritz, Robin (2000) A new, simple method for rearing diploid drones in the honeybee (Apis mellifera L.). *Apidologie,* 31, 525-30.

Bienefeld, K (2016) Breeding Success or Genetic Diversity in Honey Bees? *Bee World,* 93, 40-4. [http://dx.doi.org/10.1080/0005772X.2016.1227547]

Camazine, S, Cakmak, I, Cramp, K, Fisher, J, Frazier, M & Rozo, A (1998) How healthy are commercially-produced US honey bee queens? *Am Bee J,* 138, 677-80.

Châline, N, Ratnieks, FLW, Raine, NE, Badcock, NS & Burke, T (2004) Non-lethal sampling of honey bee, *Apis mellifera*, DNA using wing tips. *Apidologie (Celle),* 35, 311-8.

[http://dx.doi.org/10.1051/apido:2004015]

Cobey, SW (2007) Comparison studies of instrumentally inseminated and naturally mated honey bee queens and factors affecting their performance. *Apidologie (Celle),* 38, 390-410.
[http://dx.doi.org/10.1051/apido:2007029]

Di Pasquale, G, Salignon, M, Le Conte, Y, Belzunces, LP, Decourtye, A & Kretzschmar, A (2013) Influence of pollen nutrition on honey bee health: Do pollen quality and diversity matter? *Public Library of Science One,* 8, 1-13.
[http://dx.doi.org/10.1371/journal.pone.0072016]

Gençer, HV & Kahya, Y (2011) Are sperm traits of drones (*Apis mellifera* L.) from laying worker colonies noteworthy? *J Apic Res,* 50, 130-7.
[http://dx.doi.org/10.3896/IBRA.1.50.2.04]

Petersen, GEL, Fennessy, PF, Van Stijn, TC, Clarke, SM, Dodds, KG & Dearden, PK (2020) Genotyping-b--sequencing of pooled drone DNA for the management of living honeybee (Apis mellifera) queens in commercial beekeeping operations in New Zealand. *Apidologie (Celle),* 51, 545-56.
[http://dx.doi.org/10.1007/s13592-020-00741-w]

(1949) Mikrochemische Futtersaftuntersuchungen und die Frage der Koniginnenentstehung. Hess. *Biene,* 85, 89-92.

Gregory, PG & Rinderer, TE (2004) Non-destructive sources of DNA used to genotype honey bee (*Apis mellifera*) queens. *Entomol Exp Appl,* 111, 173-7.
[http://dx.doi.org/10.1111/j.0013-8703.2004.00164.x]

Harbo, JR & Williams, JL (1987) Effect of above-freezing temperatures on temporary storage of honeybee spermatozoa. *J Apic Res,* 26, 53-5.
[http://dx.doi.org/10.1080/00218839.1987.11100735]

(1957) The food of the drone larvae. Ann. ent. *Soc Am,* 50, 73-5.

Henryon, M, Liu, H, Berg, P, Su, G, Nielsen, HM, Gebregiwergis, GT & Sørensen, AC (2019) Pedigree relationships to control inbreeding in optimum-contribution selection realise more genetic gain than genomic relationships. *Genet Sel Evol,* 51, 39.
[http://dx.doi.org/10.1186/s12711-019-0475-5] [PMID: 31286868]

(1969) *J Apic Res,* 8, 65-74.
[http://dx.doi.org/10.1080/00218839.1969.11100220]

(1959)

Takeuchi, K, Watabe, N & Matsuka, M (1972) Rearing Drone Honeybees in an Incubator. *J Apic Res,* 11, 147-51.
[http://dx.doi.org/10.1080/00218839.1972.11099715]

(1972) NAOHISA WATABE AND MtTSUO MATSUKA. REARING DRONE HONEY BEES IN AN INCUBATOR. *J Apic Res,* 11, 147-51.
[http://dx.doi.org/10.1080/00218839.1972.11099715]

Koeniger, N, Koeniger, G & Pechhacker, H (2005) The nearer the better? Drones (Apis mellifera) prefer nearer drone congregation areas. *Insectes Soc,* 52, 31-5.
[http://dx.doi.org/10.1007/s00040-004-0763-z]

Koeniger, G (1990) The role of the mating sign in honey bees, *Apis mellifera* L.: does it hinder or promote multiple mating? *Anim Behav,* 39, 444-9.
[http://dx.doi.org/10.1016/S0003-3472(05)80407-5]

Laidlaw, HH, Jr & Page, RE, Jr (1984) Polyandry in honey bees (*Apis mellifera* L.) sperm utilisation and intracolony genetic relationships. *Genetics,* 108, 985-97.
[http://dx.doi.org/10.1093/genetics/108.4.985] [PMID: 17246245]

Locke, SJ & Peng, YS (1993) The effects of drone age, semen storage and contamination on semen quality in

the honey bee (Apis mellifera). *Physiol Entomol,* 18, 144-8.
[http://dx.doi.org/10.1111/j.1365-3032.1993.tb00461.x]

Composition of the food of drone larvae at different ages. J. a pic. *Res* in press

Meuwissen, THE (1997) Maximising the selection response with a predefined rate of inbreeding. *J Anim Sci,* 75, 934-40.
[http://dx.doi.org/10.2527/1997.754934x] [PMID: 9110204]

Nur, Z, Seven-Cakmak, S, Ustuner, B, Cakmak, I, Erturk, M, Abramson, CI, Sağirkaya, H & Soylu, MK (2012) The use of the hypo-osmotic swelling test, water test, and supravital staining in the evaluation of drone sperm. *Apidologie (Celle),* 43, 31-8.
[http://dx.doi.org/10.1007/s13592-011-0073-1]

(1951) Uber die Erniihrung der Drohnenmaden. *Z Bienenforsch,* 1, 63-6.

Rhodes, JW (2008) Semen production in drone honeybees [online]. Publication No. 08/130. Rural Industries Research and Development Corporation, Barton, Australia. Avilable From: https://rirdc.infoservices.com

Rhodes, JW, Harden, S, Spooner-Hart, R, Anderson, DL & Wheen, G (2011) Effects of age, season and genetics on semen and sperm production in *Apis mellifera* drones. *Apidologie (Celle),* 42, 29-38.
[http://dx.doi.org/10.1051/apido/2010026]

Roberts, WC & Mackensen, O (1951) Breeding improved honey bees. *Am Bee J,* 91, 473-5.

Rousseau, A, Fournier, V & Giovenazzo, P (2015) *Apis mellifera* (Hymenoptera: Apidae) drone sperm quality in relation to age, genetic line, and time of breeding. *Can Entomol,* 147, 702-11.
[http://dx.doi.org/10.4039/tce.2015.12]

Rousseau, A & Giovenazzo, P (2016) Optimising drone fertility with spring nutritional supplements to honey bee (Hymenoptera: Apidae) colonies. *J Econ Entomol,* 109, 1009-14.
[http://dx.doi.org/10.1093/jee/tow056] [PMID: 27018435]

Schlüns, H, Moritz, RFA, Neumann, P, Kryger, P & Koeniger, G (2005) Multiple nuptial flights, sperm transfer and the evolution of extreme polyandry in honeybee queens. *Anim Behav,* 70, 125-31.
[http://dx.doi.org/10.1016/j.anbehav.2004.11.005]

Shuel, RW & Dixon, SE (1968) The importance of sugar for the pupation of the worker honeybee. *J Apic Res,* 7, 109-12.
[http://dx.doi.org/10.1080/00218839.1968.11100199]

Stürup, M, Baer-Imhoof, B, Nash, DR, Boomsma, JJ & Baer, B (2013) When every sperm counts: factors affecting male fertility in the honeybee *Apis mellifera. Behav Ecol,* 24, 1192-8.
[http://dx.doi.org/10.1093/beheco/art049]

Tarpy, DR & Page, RE, Jr (2000) No behavioural control over the mating frequency in queen honey bees (*Apis mellifera* L.): implications for the evolution of extreme polyandry. *Am Nat,* 155, 820-7.
[http://dx.doi.org/10.1086/303358] [PMID: 10805647]

Tarpy, DR, Keller, JJ, Caren, JR & Delaney, DA (2012) Assessing the mating 'health' of commercial honey bee queens. *J Econ Entomol,* 105, 20-5.
[http://dx.doi.org/10.1603/EC11276] [PMID: 22420250]

vanEngelsdorp, D & Meisner, MD (2010) A historical review of managed honey bee populations in Europe and the United States and the factors that may affect them. Journal of Invertebrate Pathology, 103: S80–S95.

Woyke (1963b) What happens to diploid drone larvae in a honeybee colony? *J Apic Res,* 2, 73-5.

Woyke, J (1963) What happens to diploid drone larvae in a honeybee colony? *J apic Res,* 2, 73-5.

Woyke, J (1965) Do honeybees eat diploid drone larvae because they are in worker cells? *J Aplc Res,* 4, 65-70.

Woyke, J (2010) Rearing diploid drone larvae in queen cells in a colony. *J apic Rcs,* 4, 143-8.

Woyke, J (2010) Drone larvae from fertilised eggs of the honeybee *F Apic Res,* 2, 19-24.

Woyke, J (1963) Woyke (l963c) Rearing and viability of diploid drone larvae. *J Apic Res*, 2, 77-84.
[http://dx.doi.org/10.1080/00218839.1963.11100064]

Woyke (l965a) Genetic proof of drones' origin from honeybee fertilised eggs. *J Apic Res*, 4, 7-11.
[http://dx.doi.org/10.1080/00218839.1965.11100095]

Woyke, J (1965) Study on the comparative viability of diploid and haploid larval drone honeybees. *J Apic Res*, 4, 12-6.
[http://dx.doi.org/10.1080/00218839.1965.11100096]

Woyke, J (1963) Rearing and viability of diploid drone larvae. *J apic Res*, 2, 77-84.

(1962) Hatchability of "lethal" eggs in a two sex-allele fraternity of honeybees. *J Apic Res*, 6-13.

Woyke, J (1962) Natural and artificial insemination of queen bees. *Bee World*, 43, 21-5.
[http://dx.doi.org/10.1080/0005772X.1962.11096922]

(1963) Method of rearing honey bee queens and diploid drones from eggs outside the hive. *Pszczel Zesz Nauk*, 7, 63-80. a

Woyke, J & Jasinski, Z (1978) Influence of age of drones on results of instrumental insemination of honeybee queens. *Apidologie (Celle)*, 9, 203-12.
[http://dx.doi.org/10.1051/apido:19780304]

Woyke, J & Jasiński, Z (1978) Influence of age of drones on the results of instrumental insemination of honeybee queens. *Apidologie (Celle)*, 9, 203-12.
[http://dx.doi.org/10.1051/apido:19780304]

Zaitoun, S, Al-Majeed Al-Ghzawi, A & Kridli, R (2009) Monthly changes in various drone characteristics of *Apis mellifera ligustica* and *Apis mellifera syriaca*. *Entomol Sci*, 12, 208-14.
[http://dx.doi.org/10.1111/j.1479-8298.2009.00324.x]

SUBJECT INDEX

A

Acid(s) 37, 54, 56
   aspartic 37
   decenoic 37, 54
   docosanoic 56
   glutamic 37
   nicotinic 37
   oleic 37
   pantothenic 37
Activity 33, 39, 40, 60, 72
   digestive enzyme 40
   proteolytic 39
Adenylate kinase 33, 34
Agricultural pollination 2
Agrochemical cocktails 33
Allelic diversity 32, 33
   reduced 33
Amino acids 37
   exogenous 37
ANOVA analysis 55
Apicultural production system 24
Artificial 66, 87, 89
   drone honey bee production 89
   insemination 66, 87

B

Bee(s) 6, 9, 55, 68
   behaviour 68
   flying 68
   hive 55, 68
Biosynthetic pathways 55

C

Carbon copy 4
Cellular proliferation 33
Chromosomes 2
Circadian rhythm influences 6
Colonial productivity 2, 53

honey bee 2
Colonies influence 43
Colony 2, 53, 67
   disease 2
   resource consumption 53
   size, honey bee 67

D

Damage, oxidative 33
Detection of pheromones 57
Detoxification 33, 34
Developing 31, 71
   drones 31
   sperms 71
Developmental synchronicity 1, 2, 28
Diploid 1, 2, 3, 4, 10, 18, 20, 28, 35, 84, 85, 86, 87
   drones 1, 2, 3, 4, 10, 18, 20, 28, 35, 84, 85, 86, 87
   spermatozoa 3
DNA 3, 33, 86, 87
   damage 33
   microsatellites 86, 87
Drone(s) 5, 18, 19, 28, 29, 35, 37, 39, 40, 42, 44, 54, 59, 60, 68, 70, 72, 88
   caste 18, 39
   drone interaction 54
   egg 5, 18, 28, 35, 40
   fly 59, 60, 68
   healthy 19, 29, 44
   honeybee 88
   immature 54
   influences queens 42
   larval growth 37
   mating 68
   population 5
   semen volume 70
   senescence 70
   spermatozoa 72
   viability 72
Drone fertility 32, 33, 72

factors influence 32
Drone mandibular 55, 56
   gland pheromones 55
   pheromone biosynthesis 56
Drone production 1, 18, 19, 59, 73
   artificial 1
Drone honey bee 5, 57
   antennae 57
   development 5
Dwarf drones 38

E

Earth's magnetic compass 7
Ecdysis 28
Ecological conditions 18, 20, 25
Economic breeding 66
Eggs 2, 35
   hatch 2
   honey bee 35
Energy 5, 33, 37
   homeostasis regulation 33
Enzyme, antioxidant 72
Esters, fatty acid 37, 55
Eusocial hymenopteran 70
Exposure, pesticide 32
Expression 29, 33, 55
   gene 55
   immune protein 33
   transcriptional 29

F

Fatty acid 53, 55, 57, 60
   derived semiochemicals 57
   methyl-branched 53, 55
Fertilize queens 19
Fertilized eggs 1, 2, 3, 18, 28, 29, 35
Fly, virgin females 59

G

Genetic 24, 65, 68, 71, 73, 87
   constitution 71
   correlation 24
   diversity 68, 87
   variations 65, 73
Genomic 30, 44, 87
   contents 30, 44

information 87
Genotyping-by-sequencing (GBS) 87
Gland(s) 8, 31, 36, 67
   accessory sex 8
   cells, cornual 67
   hypopharyngeal 31, 36
Glomeruli, isomorphic 57
Glycogen accumulation 36
Growth, healthy 31

H

Haemolymph 39
Haploid 1, 3, 4, 18, 20, 28, 35, 38, 65, 86, 87
   drones 3, 18, 28, 35, 38, 65, 86
   inheritance 28
Haploid eggs 2, 3, 18, 29, 40
   unfertilized 3, 18
Hatching eggs 86
Helium balloon 67
Hemolymph 32, 33, 38, 39
Holometabolous 28
Homozygosity 28
Homozygous conditions 87
Honey 3, 7
   processing 3
   productivity 7
Honey bee(s) 2, 5, 6, 7, 9, 10, 19, 20, 28, 29, 41, 42, 53, 55, 65, 66, 67, 68, 70, 71, 72, 73, 84, 87, 89
   breeding programs 87
   colonies 2, 5, 10, 28, 29, 65, 66, 67, 68, 70, 71, 72, 73, 84, 87, 89
Humidity, relative 6, 85, 86
Hymenopteran 4, 28
   drones 4
   insect 28

I

Infection, reduced 31
Influence 33, 59, 69
   drone development 33, 69
   flyways 59

L

Lipid content 36

Subject Index

M

Macroglomeruli, voluminous 57
Mandibular glands 31, 56, 57, 60
Mated queens 25, 29, 60, 88
Mating 9, 19, 56, 59, 67, 68, 69, 71, 88
    flights 9, 19, 56, 59, 67, 88
    health 88
    of honey bee queens 67
    procedure 71
    sign 67, 68, 69, 71
    success, sperm quality influence queen 88
Mechanical stimulation effects 39
Mechanism 2, 65
    haplodiploid sex-determination 2
Meiotic gametogenesis 4
Mellifera biotypes 73
Metabolic 55, 72
    capacity 72
    pathways 55
Metamorphosis 36
Methyl oleate (MO) 55, 60

N

Nurse bees 1, 31, 42, 44, 53, 54, 56
    healthy 31
Nutrition, insect 84

O

Organogenesis 29, 38
Ovarian development 18
Oxidative metabolism 39

P

Pesticide(s) 32, 33
    clothianidin 33
    treatment 33
Phenotypic correlation 24
Photoisomerization 60
Pollen 5, 10, 19, 31, 36, 38, 39, 40, 44, 72, 89
    consumption 40
    fresh 31
    grains 5, 10
    supplements 40
    supply 31, 44

Polyandrous 9, 18, 88
    mating system 88
Polyandrous queen 28, 65
    honey bee mates 28
Post 2, 33
    mating death 2
    translational modification 33
Predation influences 40
Productivity 28, 29, 65, 73, 84
    bee population influences 73
Protein(s) 25, 31, 32, 33, 34, 36, 37, 38, 70, 72
    amino acid storage 34
    composition 70
    formation 25
    rich diet 31
Proteomics 29
Putative glutathione-S-transferase 34

Q

Queen 1, 2, 9, 10, 19, 29, 35, 40, 59, 60, 65, 67, 71, 73, 87
    bee mating 29
    honey bee mates 1, 19, 71
    mandibular gland 60
    mates 2, 9, 10, 35, 40, 65, 73, 87
    mating 19, 67
    sex pheromones 59, 67
Queenless condition 6

R

Reproduction, sexual 1
Reproductive system 1, 8, 18, 65
Riboflavin 37
Ripe honey cells 3
Ruttner's method 23

S

Salivary gland 55
Season 10, 19, 35, 43, 66, 69, 70, 71, 88
    queen-rearing 88
    reproductive 66, 69
Secretions 31, 36, 60, 71
    gland 31
    queen mandibular gland 60
Semen 72, 89
    production 89

quality 72
Seminal 9, 21, 23, 24, 70, 71, 72, 89
    fluid 70, 71, 72
    vesicle (SV) 9, 21, 23, 24, 70, 71, 89
Sensory system 57
Signals, vibration 69
Single-nucleotide polymorphism (SNPs) 87
Sperm(s) 1, 8, 10, 24, 25, 29, 30, 42, 43, 65, 66, 67, 69, 70, 71, 72, 88, 89
    cells 8, 66, 70, 72
    competition 72
    development 69
    injection 67
    migration of 70, 71, 89
    production 25, 30
    queen stores 29
    stored 30, 88
    viability 10, 25, 70, 72, 88
    viscosity 10, 70
    volume 24
Spermatogenesis 66
Stimulate oogenesis 39
Stress 32, 33, 34, 72
   cold 32, 34
   heat 34
   oxidative 34, 72
   pesticide 33
   response, putative 32
   tolerance 32
Stressors 32, 33
   abiotic 32
   biotic 33
   environmental 32, 33
Stress response proteins 33, 34
   putative 33, 34
Superoxide dismutase 72
Synthesis, eclosion juvenile hormone 39

T

Testicular follicles 8
Triglycerides 37
Trophallaxis 54, 69

W

Wind(s) 43, 59
   high 43
   velocity 59

Wing morphology 71

X

Xenobiotics 32